Bourbon Street, B-Drinking, and the Sexual Economy of Tourism

Anthropology of Tourism: Heritage, Mobility, and Society

Series Editors

Mission Statement

The Anthropology of Tourism: Heritage, Mobility, and Society series provides anthropologists and others in the social sciences and humanities with cutting-edge and engaging research on the culture(s) of tourism. This series embraces anthropology's holistic and comprehensive approach to scholarship and is sensitive to the complex diversity of human expression. Books in this series particularly examine tourism's relationship with cultural heritage and mobility and its impact on society. Contributions are transdisciplinary in nature and either look at a particular country, region, or population or take a more global approach. Including monographs and edited collections, this series is a valuable resource to scholars and students alike who are interested in the various manifestations of tourism and its role as the world's largest and fastest-growing source of socio-cultural and economic activity.

Advisory Board Members

Recent Books in Series

Bourbon Street, B-Drinking, and the Sexual Economy of Tourism, by Angela R. Demovic
Anthropology of Tourism in Central and Eastern Europe: Bridging Worlds, edited by Sabina Owsianowska and Magdalena Banaskiewicz
Apprenticeship Pilgrimage: Developing Expertise through Travel and Training, by Lauren M. Griffith and Jonathan S. Marion
Cosmopolitanism and Tourism: Rethinking Theory and Practice, edited by Robert Shepherd

Bourbon Street, B-Drinking, and the Sexual Economy of Tourism

Angela R. Demovic

LEXINGTON BOOKS
Lanham • Boulder • New York • London

Published by Lexington Books
An imprint of The Rowman & Littlefield Publishing Group, Inc.
4501 Forbes Boulevard, Suite 200, Lanham, Maryland 20706
www.rowman.com

Unit A, Whitacre Mews, 26-34 Stannary Street, London SE11 4AB

British Library Cataloguing in Publication Information Available

Library of Congress Cataloging-in-Publication Data

Library of Congress Control Number: 2018940325 ISBN: 978-1-4985-3132-0 (cloth) ISBN: 978-1-4985-3133-7 (electronic)

∞™ The paper used in this publication meets the minimum requirements of American National Standard for Information Sciences Permanence of Paper for Printed Library Materials, ANSI/NISO Z39.48-1992.

Printed in the United States of America

For my Jaanu

Entering "The Field"

A Preface

In August 1990, I began graduate school in New Orleans to study psychology but soon shifted my focus to cultural anthropology. The region was geographically and culturally novel to me, having grown up in Chicago and its suburbs. The cultural icons of French and Spanish colonialism, the range in architecture from simple cottages to the ornate ironwork balconies, the divisions on the landscape where the Creoles' and later-arriving Americans' styles of life and housing collided—together these illustrated stories of a past and an aesthetic different from the America I knew. I read tourism pamphlets while drinking the one-drink minimum and sitting in the bars of historic hotels which had not only brochures but free food during happy hour, a sanctuary for the poor graduate student. Many of those buildings had staff who were trained to tell me something about the antiquity of the building or, sometimes, more famous adjoining buildings. I was fascinated by the French Quarter and spent many hours investigating the place and its history. I was in the thrall of the city, and I continue to be today.

I pursued and obtained the Tour Guide Permit through testing overseen by the Taxi Cab and For Hire Bureau, which consisted of some very basic facts: the site was chosen by Jean-Baptiste Le Moyne de Bienville in 1718; the city was a colony of the French, the Spanish, and then the French again; Nouvelle Orleans, now the French Quarter, was laid out in a grid with Jackson Square facing the river at its center; St. Louis Cathedral is in the original spot but has been rebuilt following two fires; and so on. I spent a few weeks shadowing tour guides on their tours, a part of the training each tour company requires, and realized the historic day tours were my forte rather than the evening ghost tours. During the summer months, between semesters, I developed and provided my own two-hour guided walking tours describing the history and creation of the French Quarter and Garden District to tourists and friends visiting the city. I developed short stories about how material objects reflected the culture, history, and language of the city. I walked backwards, leading tourists between significant sites as I spoke, describing hidden courtyards set back from the dust and smells that would have (and still do in some sections) infused the *banquette*—today the sidewalk—which was initially raised above what would have been dirty, wet streets in a newly developing colony. I pointed out examples of houses with no or very little front yard space. To develop my tour, I poured over increasingly reliable texts about the history of the city in Tulane's multiple libraries and created an elaborate pocket notebook that I carried and used as a reference. Time spent chatting with other tour guides taught me additional details about buildings and monuments, but more importantly clarified for me the constructed nature of even the historic tours, and the many stakeholders and processes involved in those constructions.

Tour guides were armed with a handful of official facts, but the needs of tourists who are enjoying a leisure activity with an educational component went beyond an historic map of the landscape. A two-hour walking tour requires stops at even and frequent intervals, not only to rest and listen but also to cool off. Favorite sites had a story and someplace to buy a drink, whether lemonade or Mint Julep. The sites that city officials at the Taxi Cab Bureau had examined me on, central to the city's authorized image of New Orleans, did not always lend themselves to frequent enough stops. On a bad tour, tired tourists walking a block or more between sites wondered aloud when the next stop might be, where the next bathroom was, and where they could purchase their next drink. They were engaging in heritage tourism, but they were generally less concerned with historical accuracy when it was disruptive to leisure and pleasure in the southern heat. Tour guides depended on tips, and thus successful guides were cautious to balance the heritage tourists' dual interests. Heritage tourists want to know about moments of history and practices, but they want to do so within the frame of tourism, which is a frame of leisure and pleasure.

These stories of the French Quarter are an important aspect of "intangible heritage," nonmaterial aspects of the heritage tourism experience that are nonetheless critical to its construction. This commercialization of history inspired tour guides to unearth, relocate, or sometimes outright invent "historic" stories for spaces that landed midway between buildings or environmental features separated by too far a distance for a comfortable walk in the subtropical humidity and sudden downpours that characterize the city's climate. Earning an income telling the history of New Orleans to visitors required keeping the interest of contemporary audiences. The constructed nature of the stories was at first more obvious to me in the performance of guides. Walking tours of the French Quarter and the bars and coffee shops that hosted them were a parallel business in many ways to Bourbon Street clubs, though beyond the scope of this book. Tour guide licensing is more complex today than it was in the 1990s.

2. B-drinking is firmly within the range of "sex work," the term adopted by academics, NGOs, and sex workers' rights communities writing about those whose jobs include a sexual service component. The term was first coined by sex worker and activist Carol Leigh and entered common use with the 1987 publication of *Sex Work: Writings by Women in the Sex Industry*, edited by Frédérique Delacoste and Priscilla Alexander. Sex work includes a broad spectrum of jobs. Some sex work is legal (such as striptease and private dancing) and some is not (such as lap dancing, prostitution, and B-drinking). Sex work may or may not include physical contact, such as phone or internet operators of sexually oriented services. Similarly, B-girls are paid to engage in sexually oriented talk and entertainment of customers, and some earn a steady income with very little physical contact with customers.

3. The names of individuals and clubs used in this book are pseudonyms to protect the anonymity of my research participants. Although dancers regularly choose a stage name, in my work I have chosen to change even those names to ensure their privacy.

Acknowledgments

I would like to thank my colleague Deborah Gordon, of the Wichita State University Women's Studies Program, for listening to my ideas as the book developed and for providing professional guidance.

I also appreciate the amazing students who helped me to gather archival data, sort articles, and proofread endless drafts. Rachael Sebastian, Elizabeth Lonning, Audrey Brown, Sapphire Garcia, William Silcott, Emily Jones, and Summer Steenberg—thank you for your time and your hard work.

For the financial support that enabled me to include my students in collecting archival data in New Orleans, I thank the Jackman Foundation for Anthropological Research.

I am grateful to the series editors, Michael A. Di Giovine and Noel B. Salazar for their reviews of the earliest drafts of the manuscript. Thanks also to the anonymous reviewer whose page-by-page critiques and comments challenged me. I also appreciate the hard work of Kasey Beduhn and Becca Rohde at Lexington Books.

Many thanks to Irene Wainwright and Wayne Everard of the City Archives at the New Orleans Public Library for their assistance in gathering newspaper articles and other historic documents. To Susan Tucker and Chloe Raub, heads of the Newcomb Archives and Vorhoff Library at the Newcomb College Institute of Tulane University during this project, thank you for your hospitality in allowing my students and I access to the collections. Thanks also go to the librarians at the Law Library of Louisiana and the Tulane Law School Library for assistance gathering relevant ordinances, legislative records, and appellate documents.

My late father, Edward S. Demovic Jr., was always skeptical about anthropology but was unfailing in his support of my studies, research, and travels as an anthropologist. He much preferred stories about life in the

French Quarter to my stories about faraway places, and he encouraged me to write the ethnographic section of this book. I am grateful to him and my mother, Charlene, for their belief in education and for their unending support.

Thank you to my husband, Ikram, who has not read this manuscript but made it possible for me to write. I am grateful for his patience, for his taking over unfinished household tasks, and for caring for our family through all my late nights of writing.

Finally, thank you to the women who contributed to the book by sharing their stories, their photos, their dreams, and their dressing rooms with me.

Introduction

In 1949, on the Sunday before Mardi Gras (Fat Tuesday, the day before Ash Wednesday and the start of Lent in the Catholic liturgical year) the *Times-Picayune New Orleans States* newspaper ran an article in the society section about the pending arrival of the fictitious King Rex by boat on the riverfront. Rex, King of Mardi Gras, is quoted as reciting several verses of his self-proclaimed national anthem "If Ever I Cease to Love." His urging the public to remember the lyrics allows the staff writer to reprint them, re-creating a part of the elaborate performances of the season of Carnival that lead up to Mardi Gras in New Orleans. This careful crafting of the bawdy society pages of New Orleans' most widely circulated newspaper on the last and most elaborate Sunday of the Carnival season also included an artful description of the New Orleans B-girl.

B-girls are integral to the French Quarter and its fantastic performances of whimsy, irony, romance, and sexuality. B-girls are stock characters in many different narratives about the city, as I will demonstrate. But what is a B-girl? Below, a mid-century staff author (also from the *Picayune*) allows us to be a fly on the wall for a conversation between wealthy revelers who are enjoying the Carnival season that Sunday in the style of New Orleans aristocracy. Masquerading as royalty, a visitor to the city wonders aloud, "When do the B-girls come on board?" His friend corrects him about the nature of B-girls and place, thus initiating a conversation about the role of such women as seductive conductors in the flow of carefree vice within the bars of the city.

The seasonal public performances of drunkenness and merrymaking by elite Carnival clubs (called Krewes in New Orleans) provides the backdrop to their fictive conversation, which could just as easily have taken place fifty years later. They reference New Orleans with the title Crescent City, a nickname still used today referring to the bend in the Mississippi that embraces the oldest parts of the metropolitan area.

> "Well," aid [*sic*] the Khalif, "to get to this great Crescent City by sea, you must be piloted across the bar. A B-girl is a bar pilot. She pilots you to the bar, across the bar and to a table if you are not careful."
>
> "Is that so?" said the unwitting Khan. "What a strange custom. But why do they call them B-girls? Is it because they work in grade B places, or because they are as busy as bees (and you might get stung) or merely because they operate on Bourbon Street?"
>
> "The exact answer to that I cannot tell you," whispered Ed Soul, "but I will say that on one of my previous voyages to our winter capital a B-girl put the bee on me for 50 drachmas and in every case on record the system is roughly you get one drink for the price of two instead of two drinks for the price of one. Still they are all human B-ings."
>
> "In fact," continued the Khalif, "I seem to recall a highly educated Ph-B-girl named Bodacious Bibulous Betsy who use to declaim sadly into her B-eer as follows: To Be or not to B, that is the question. Whether 'tis sober of my kind to suffer the (gin) slings and arrows of a butter and egg fortune or take alms against a sea of bubbles (champagne to you). Ah, there's the sub(statute) for which I earn my pay.'" (*Times-Picayune*, Feb. 27, 1949)

Although this article was written in the late 1940s, B-drinking as a modality of sexual commerce (Frohlick 2013, 157) continued to organize highly gendered and erotic encounters between female and transgender[1] service workers and their customers—visitors to the New Orleans' French Quarter—until at least the end of the twentieth century. The services of B-drinkers have been desired by well-heeled travelers and have been provided by mostly feminine people in New Orleans' French Quarter since at least the late nineteenth century. Seemingly little changed between the 1940s and the late 1990s in how this marginalized work was organized. It has remained both a stigmatized form of labor and an economic boon to B-drinkers, other nightclub employees, and club owners alike.

I borrow the terms *B-drinking* and *B-girl* from the Louisiana legislature, where I find they have been defined and defended in writing between the mid- and late twentieth century. B-drinking is illegal, prohibited by still largely unchanged acts of the Regular Session of the Louisiana legislature in 1948 under the Earl K. Long governorship. In practice, it is the act of a woman (usually a bartender, waitress, or exotic dancer) soliciting (or having solicited by a third party) a club patron to order a drink for both himself and his interlocutor. She is legally a B-girl because she is paid at the end of her working shift, and her income is earned on a per-drink basis. B-drinkers (the nondiscriminatory term that enters the law in 1985 after decades of appeals, as we will learn in chapter 3) are paid on commission to sell and drink as companions to club patrons in an erotic and ludic environment, a form of emotional labor (Hochschild 1983) like an in-house escort.

Outside of the city of New Orleans, B-drinking goes unmentioned in literature on the intersectionality of intimacy, alcohol, and tourism encounters. My research reveals evidence that the practice earned high incomes for many entertainers and service-industry workers in New Orleans during the twentieth century and into the start of the twenty-first century. It is B-drinking and the persistent (re)creation of B-girls as a categorical type of character that inhabits the French Quarter, particularly the strip clubs on Bourbon Street, that are the subject of this book.

Buying a B-drink is a sort of contract for the time, conversation, and flirtation of the B-drinker. It is this performance of sexuality that animates the French Quarter's reputation as a tourism site where adult entertainment is not only available, but where images and performances of sexuality have long been a part of the construction of place that is marketed to visitors. B-drinkers engage in emotional labor, controlling their public display of emotion and performing while both on and off the stage of strip clubs. To the delight of the visitor, this commodified engagement with the Other characterizes the nightclubs of Bourbon Street, a tourist site where heritage and sexuality intersect in understandable and permissible ways. As in other tourist destinations, a master narrative (Bruner and Kirshenblatt-Gimblett 1994) is performed, one which is about gender, display, and availability. It is a drama that produces what Edward Bruner and Barbara Kirshenblatt-Gimblett called tourism realism: "an effect closely linked to the ultimate tourism commodity—experience" (435).

As an ethnographer, my interview data revealed to me that at the turn of the century B-drinking was economically central to the range of erotic performance expected by and sold to tourists of Bourbon Street nightclubs, despite the more visually apparent performances of semi-nude dance. B-drinkers provided tourism services in response to the desire of the tourist to engage with authentic New Orleans' people in an intimate way. Although other studies of strip clubs have focused on relationships with regular customers, strip-club interactions on Bourbon Street are mainly between contract dancers and visiting tourists. The highly sexualized roles found within the American strip club are performed on Bourbon Street in a nuanced way, both because of the notorious history of the city and because, unlike other strip clubs around the country, the clubs on Bourbon Street are almost wholly dependent on tourists for their incomes. Small clubs and their employees struggled during the summer tourism slump, when conferences and tourists avoided the sultry heat of the southern wetlands.

In the ethnographic chapters of my book I present the voices of B-drinkers who were waitresses, bartenders, or who performed as striptease dancers between 1990 and 2005. B-drinking was a requirement for workers in these otherwise legal jobs in many Bourbon Street strip clubs between mid-century and at least 2005.[2] Despite being vulnerable to stigma and arrest, B-drinkers in the 1990s were powerful stakeholders in the French Quarter, particularly, but not exclusively, within the Bourbon Street cluster of nightclubs. Although the working conditions of the clubs dependent on B-drinking often resulted in stigma and danger for B-drinkers, their potential incomes were high enough not only for survival but also to save money. B-drinkers earned enough to save the necessary capital to invest in their own or their children's educations, in small businesses, or otherwise. The opportunities for lucrative erotic labor on Bourbon Street drew women from not just the Greater New Orleans area but from all over the United States to accept such work and its conditions. The risk, at least in part, stems from a legal system that preys upon their continued structural positioning as criminals. A careful look at B-drinking and its role in the intimate markets of the French Quarter provides an intersection for tourism studies, anthropology, and feminist jurisprudence.[3]

ETHNOGRAPHIC METHODS: LEARNING FROM B-GIRLS FIRSTHAND

I first began my notetaking on strip clubs as a graduate student taking a class on research methods; I did interviews, photo studies, led focus groups, and practiced other methods used by anthropologists in cooperation with a group of about 25 women working at one nightclub. These first interviews and my participant observation in the lives of women working on and around Bourbon Street occurred between 1992 and 1997, picking back up between 1998 and 2001. During these years, I lived in New Orleans and spent three or four nights of each week observing and informally interacting with people engaged across the range of formal occupations on Bourbon Street, especially those that included B-drinking. Thus, my analysis of B-drinking as a practice is based on participant observation in French Quarter strip clubs at the turn of the century (1990–2005).

I visited French Quarter nightclubs, restaurants, coffee shops, adult-oriented clothing stores, and other tourism-focused workplaces throughout the 1990s taking notes and performing informal interviews. My interviews eventually expanded to include more than 70 women who worked in three French Quarter topless dance bars, two on Bourbon Street and the third a block away. These nightclubs were places where B-drinking was an expected practice and made up a significant portion of the earnings of nightclub employees (Demovic 1993). While my own observations of the practice occurred in strip clubs frequented mostly by tourists and business travelers, interviews revealed that the strategy is not limited to such clubs, nor did it occur in all such clubs.

Picking up my interview work again from 2012 until 2016, I spent summer months seeking out detailed, semi-structured interviews with some of these same women, contributing to a more longitudinal understanding of how women experience this type of sexualized labor. In these most recent interviews, women who had long left the stage reflected on their experiences while employed on Bourbon Street, remembering their personal experiences and how they had affected their later lives. My participants also told me about the range of places their careers as dancers and B-drinkers had taken them, suggesting that earning an income soliciting patrons of strip clubs for drinks played an important role not only in New Orleans, but more broadly in the sexual economy of the Gulf South.[4]

IS THERE NO SEX IN THE CHAMPAGNE ROOM?

In a previous study of Bourbon Street, Vesey and Dimanche assert, "One of the most alluring features of the city, in its American context, is its history of vice, festivals and its subtropical climate" (2003, 60). Interpreting contemporary Bourbon Street as a commercialized representation of the historic Storyville, a legal red-light district that closed during the First World War, they see the space as meant to evoke nostalgia. It is that nostalgia for an imagined past that attracts tourists to the city. Tourism "plays out this emotional longing," nostalgia, "in very material terms" (Di Giovine 2009, 154). Tourists want to engage with real people. B-drinkers meet the tourists' desires to interact with local people while strategizing to build a mock rapport by representing an expected type of gendered Other, while actively minimizing the visibility of their economic goals (see Little 2003 for a similar example).

Parallel to Susan Frohlick's analysis of the transcontinental tourist in Costa Rica, the tourist in the New Orleans strip club seeks to find particular "economically ambiguous sexual and often intimate relations" which are expected to reify (or make real for them) a particular "social sexual history" (2007, 139) of place. Vesey and Dimanche find that although "Bourbon Street and its tourism business try to convey a nostalgic image of a time when vice and sin were commonly found . . . those days are over" (2003, 66). They claim that the *image* of vice (62), the memory of Storyville, and a cultural space where tolerance for public drinking, dancing, and public nudity attracts tourists and economically pushes the development of the place.

Describing Bourbon Street, they observed

> girls dressed in their striptease outfits at the door of adult clubs. Bourbon Street provides a romantic air of when Storyville existed with its bars and strip joints with B-girls and barkers in the street competing for the many patrons looking for a unique New Orleans experience.
>
> The seeming vice activity that takes place, however, is under strict controls from the city. Strip clubs operate with controlled guidelines of how to show appreciation for a woman's dance; men may not touch the dancers, money exchanges must take place with three feet between the dancer and the patron, and no sexual activity is allowed on any level. Hence, strip clubs are only emblematic representations of open sexual expression. The rules involved actually limit sexual behavior, not promote it as Storyville once did. (2003, 63)

My longitudinal, ethnographic involvement with the women of Bourbon Street leads me to very different conclusions about the availability of commoditized sexual contact on Bourbon Street in the French Quarter. My informants made clear to me that indeed there were rules that limited sexual behavior, but they also revealed ongoing struggles between management and workers over who would set those limits and how. Power to control encounters within the strip clubs of Bourbon Street in the 1990s was not static but rather continually challenged, and I am interested in the perspectives of workers as agents of their own decisions despite the formal laws at hand.

In the 1990s B-drinking provided high incomes for dancers and bartenders in New Orleans' most heavily toured neighborhoods. It increased overall alcohol sales in clubs, increasing the incomes of not only club owners but the doormen and waitstaff. Air conditioning repair persons, cleaning persons, costume seamstresses, and many other businesses that supplied services to the clubs depended upon the incomes of such nightclubs. Their incomes were increased because of money that came into the French Quarter from tourists through B-drinkers and nightclubs that encouraged and sometimes required the practice of their feminine employees.

The B-girl caters to the desires of tourists for interactive experiences, and the measure of her commodified labor is a percentage of her drink sales. Tourists seeking various levels of intimate interactions can find them performed off the stage by the striptease dancers, with the length of time it takes to consume one drink, many drinks, or a bottle of champagne, softening the unexpected transactional aspects of such intimate relations. Customers seeking physical contact with women on Bourbon Street can negotiate their serious intent through the purchase of such expensive drinks. From the worker's perspective, B-drinking in semi-private champagne rooms provided a safe space from which to solicit big-spending customers for other forms of lucrative sex work both inside and outside the bar.

I believe the difficulty of other researchers in finding any actual sex for sale on Bourbon Street attests to the hidden, therefore safer (from arrest), nature of sex work for women who solicited customers from within the clubs as B-drinkers. Sex workers are more invested in concealing their illegal activities than academic researchers are in locating them. Sex workers are aware of the uses of law (if not always the explicit wording of it), and use interactions in spaces like the semi-private VIP rooms of strip clubs to carefully and collaboratively vet potential customers as part of their strategies to minimize risk of arrest and violence.

A short-term researcher in any club, especially a female over age 30, has little chance of finding herself the recipient of an offer for involvement in transactional sex on Bourbon Street. Even as a long-term member of the community in one club, it was only when I was explicitly invited to join two women who were doing "out-call" work after club hours that I learned how such work was organized. My rejection of the offer was initially viewed as reluctance over price, and the quiet attempts at negotiation revealed to me something about how much sex workers might earn as secondary incomes outside the club and how they protect themselves and each other. Ultimately, when my friends learned I was neither hostile to nor interested in participating in out-call "parties," they simply stopped telling me about them. Why risk it? I learned more about such work in my follow-up interviews with women decades later—as distance in time made revealing details safe.

As I will explain, B-drinking places historically notorious cultural icons into a position of structural violence. In the tradition of the Chicago School of Sociology, I investigate this segment of urban culture through the ethnographic lens. Long-term participant observation with this community combined with investigation into the history of this construction of work contributes to our understanding of the processes involved in the creation of structural inequities and increased risk of violence for women within this occupational group. The ethnographies of everyday life and "communities within the communities" written in Chicago between 1917 and 1942 and the Chicago School of Sociology and its "stepchild" (Deegan 2001, 11) the Chicago School of Ethnography, are inspirational to this text.

"THEY ARE ALL PROSTITUTES, YOU KNOW": IMAGINARIES OF BOURBON STREET

Twice in my early visits to the French Quarter to spend time in strip clubs, I was given the above warning by male friends about the strip clubs of Bourbon Street. Although neither had spent any time in the clubs (according to their claims), and neither had any kind of experience working in the tourism industry nor in the French Quarter specifically, both felt they knew what such places were like. Where did their ideas come from? The ways that destinations become collectively imagined, the processes that contribute to such imagining, and the effects of such imaginaries on social dynamics can all be illustrated with the example of the B-girl.

Maria Gravari-Barbas and Nelson Graburn refer to this phenomena as the "tourist imaginaries" of place. Such imaginaries "represent a specific sector of the overall worldview of individuals or social groups concerning places outside their primary residence where certain types of leisure activities could take place" (2012,1). The tourist imaginaries precede the travel, as the first thoughts about a potential destination lead to the development of an itinerary of activities based on multiple, shared representations of the site. These representations are borne out through multiple material images created by a broad range of stakeholders. Such "image fashioning" (Gravari-Barbas and Graburn 2012, 2) of potential leisure activities has been influenced by popular media, tour guides, and novels to include Bourbon Street as a focal point for visitors. A multiplicity of such stories situates B-girls in place. The tourist imaginary blends historic moments and stereotypes of place with contemporary realities. Over time, New Orleans has evolved as a destination characterized through an imaginary of vice and excess which includes the B-girl. It is this tourist imaginary that has led to the cultural formation of the reality of B-drinking as practice.

Michael Di Giovine discusses how the imaginaries of places are produced as a dialectical process, "presenting, imagining, re-presenting, and re-imagining" (2014, 150). Local inhabitants, tour operators, preservation specialists, and consumers all take part. Gravari-Barbas (2003) also identifies heritage as such an arena of contestation and negotiation, with the process of "heritagization" recognized as being goal oriented (Poria 2010).

In New Orleans, the imaginary of the city has come to include the personage of the B-girl. Contracted as a dancer, a waiter, or a bartender in the formal economy, between sets or during slow times she comes into clearer focus as not only the object of a tourist gaze (Urry 1990) but also an active performer who creates a multisensory experience that includes interactions that lead the tourist to experience a change in arousal state. John Urry's notion of the tourist gaze (1990) prioritized the visual, but the garish display of hinted sexual services and erotic experiences within the clubs of Bourbon Street include smells, touch, and sound as key components of a red-light sensorium which reinforces tourist imaginaries.

Strip clubs on Bourbon Street, like Bruner and Kirshenblatt-Gimblett's (1994) example of Mayers Ranch in Kenya, re-create a "master narrative." This master narrative is performed daily to an international audience, creating a tourism reality that is produced, created, and sold as the commodity of experience. What was for sale at the ranch in Kenya and in the nightclubs on Bourbon Street both included a "sentiment," an emotive experience invoked by the performance and sold as a tourist production. As Mayers Ranch in Kenya staged the colonial fantasy of the noble savage metaphorically constrained by colonialism's British garden, Bourbon Street stages a gendered fantasy of the constraint and dangers of women.

The historic evidence presented here tells more of how New Orleanians thought about B-girldom than the tourist, as it relies mainly on local news and legal documents. I place the B-girl and B-girldom as a local idea within what Naomi Leite refers to as *tourism imaginaries*.

> Tourism imaginaries are those imaginaries—conceptions, images, and imaginings of self or Other, place or people, abstract moral order or particular historical site, variously held by tourists, providers, local populations, development consultants, marketers, guides, etc.—*that are not necessarily particular to tourism, but in one way or another become culturally salient in tourism settings*. [emphasis mine] (Leite 2016, Afterword)

The B-girl predates tourism and cannot be "particular" to it. Rather, through my first chapters I follow the evidence for the production of a conception of a person, a stigmatized Other, who has held changing roles of liminal femininity across many decades and who "one way or another" (Leite, above) had become an important part of Bourbon Street, and thus an icon of French Quarter tourism, by the 1990s.

NEW ORLEANS AS A HERITAGE TOURISM SITE

Heritage is a powerful concept, "a conceptual memorialization of a specific ideological claim about the past, and a group's particular relationship to it in the present and the future" (Di Giovine 2009, 91). Kevin Fox Gotham (2007) claims that in marketing New Orleans to visitors and investors, the strategy of urban branding has been "constructed around three themes—history, music, and food," which he calls the "holy trinity of New Orleans tourism" (20). The popularity of historic walking tours, mentioned in the preface, support his claim. Part of that history, as the tour guides' story of the house of the rising Creole sons suggests, is about commodified sex.

With the decline of the shipping industry in the mid-twentieth century, the waterfront of New Orleans opened, and city planners actively pushed for the development of tourist attractions, many of which piggyback on this urban brand (Gotham 2007:13). The term "heritage" in its colloquial usage is not only of interest to anthropologists but has been a highly manipulated concept deployed in the purposeful development of the city's events. For example, the multi-stage Jazz and Heritage Festival begun in the 1960s was planned to be held in May to increase arrivals during an annual tourism lull (Sakakeeny 2013, 97). This top-down power of urban branding has contributed to increased tourism arrivals and expenditures continuing into the twenty-first century (Gotham 2007, 14–15).

Tourism as sightseeing (Urry and Larsen, 2011) can seem to revolve around the concrete, as destinations are cognitively mapped before travel through itineraries to tour immovable features of cultural patrimony. The potential to ride a paddleboat on the Mississippi River to visit the Chalmette Battlefield, site of the 1815 Battle of New Orleans, can be booked in advance online, one way the history of the riverfront becomes a destination. The potential to ride aboard the modernly equipped paddleboat, the Creole Queen, adds the authenticity of the paddlewheel itself, although the boat is a modern reconstruction of this antiquated technology, an artifact of the year 1983. The experience re-creates a dramatization of a previous era as part of staging a tourism reality (Bruner and Kirshenblatt-Gimblett 1994), "catering to the imagination of others" (Waller 1993, 301) that occurs here not in a closed destination resort or amusement park but at the "geographically specific historical sites" of the New Orleans riverfront and field of battle, mixing these two ways of learning history while traveling but "blur(ing) the distinction among them" (Waller 1993, 435).

The Creole Queen's advertising blends an explanation of what is reproduced on the trip and how that history is staged, to attract and engage the tourists.

> Constructed in Moss Point, Mississippi, the Paddlewheeler Creole Queen took her maiden voyage on October 1, 1983. She is an authentic paddlewheeler powered by a 24-foot diameter paddlewheel and the most modern, fuel efficient GE diesel-electric system ever installed in a boat of this class. Even though the Creole Queen would have been right at home among the steamboats of 1850's New Orleans, passengers of that era could never have conceived of a paddlewheeler with her luxury, comfort and safety. . . .
>
> The Creole Queen riverboat features outer decks with authentically patterned wrought iron trims inviting guests to stroll back in time as they settle into the mood of a bygone era. The interiors are richly appointed with plush Victorian style draperies, soft period lighting, wooden parquet dance floors, and Louisiana Cypress bars accented with brass railing for a warm and welcoming setting. (New Orleans Paddlewheels, Inc.)[5]

It is worth noting that the description does not focus on the visual but rather the more ephemeral feelings and senses it evokes. Passengers are invited to feel luxurious, comfortable, and safe as they "settle into the mood of a bygone era." The adjectives used—plush, soft, wooden, warm—all are evocative of the sense of touch, not sight. These are aspects of what Emma Waterton and Steve Watson refer to as the "semiotics of heritage tourism":

> the particular aspects of emotion and affect that can be sensed in the landscapes of heritage tourism and the way that these, as embodied experiences,

> can interact with the representational aspects of those same landscapes. (2014, 3)

Tangible heritage such as monuments, art galleries, and antiques offer "a literal way of touching the past" (United Nations Educational, Scientific and Cultural Organization, UNESCO website). Within the French Quarter there are many architectural sites that have been preserved as representations of New Orleans' place in American, French, and Spanish colonial history. Visitors and locals stroll through Jackson Square, often taking a photo of themselves in front of the towering St. Louis Cathedral. Rebuilt on the site of the first church in the French Louisiana Colony, a tour of the building evokes strong sentiment, and tourists can smell the incense and engage in Catholic rituals of candle lighting as a part of their experience. Recently, the discourse of physical, presentational heritage—heritage displays like memorials and statues—has risen to the level of more conscious and open discourse about how New Orleans should remember itself and be remembered, with the controversial removal of Confederate statues like General Lee, the previous occupant of Lee Circle.

Another example of tangible heritage, like a snapshot of an idealized moment in the city's history, can be found in the careful restoration and display of the mid-nineteenth-century Hermann-Grima House. It is the museality of the house—the separation of the objects from their original functions and their interpretation primarily as a collection of social, museum objects for exhibition and teaching (Mohr 2011)—that lends it particularly to sightseeing. Complete with period furnishings, a garden in the courtyard, and restored slave quarters open for tours, the home presents the history of New Orleans' elites.

But is heritage tourism only about material sites? Visitors to New Orleans are also interested in the practices of the people of the city, the varied forms of street art and ritual which are central to the experience of the Quarter, from street musicians to jazz funerals. This category of *intangible heritage* includes the performances that create the character of Bourbon Street as a place, such as the B-girls and striptease dancers of Bourbon Street that are the topic of this book.

The concept of intangible heritage, first recognized by UNESCO in 2003 includes "the practices, expressions, knowledge and skills that communities, groups and sometimes individuals recognise as part of their cultural heritage."[6] I argue performances like striptease and B-drinking in New Orleans are critical components of such intangible heritage.

Like buskers, the musicians that perform for tips in the French Quarter and whose placement is regular, logical, and mappable (Sebastian 2015), the performances that occur on Bourbon Street have come down through time as an expected and indeed definitional part of the French Quarter as a tourist destination. In a faraway comparison, the animation of Bourbon Street by performers is like the appearance of the Monday market of Djenné, Mali (Joy 2016, 112). Just as Charlotte Joy found that tourists planned their visits around the timing of the Monday market, tourists plan to visit Bourbon Street during the times when the space is animated (although the physical site is there to visit at any time). In another similarity to this faraway market, the performances of Bourbon Street are not purchased by every tourist, yet the setting and people along the strip still "serves as dramatic foreground activity in their [the tourists] photographs" (Joy 2016, 112). Like Djenné and many other heritage destinations, New Orleans is heavily defined by its intangible heritage, particularly artistic performance.

For Di Giovine, with the addition of this definition of intangible heritage "UNESCO comes full-circle, creating a cohesive heritage-scape that fuses the ephemeral and tangible expressions of culture" (2009, 116). The recognition of the value of such nonmaterial cultural patrimony as "practices"—such as the performances of B-drinkers—expands the conversation of the development of heritage tourism and its components for the tourist. More critically to this study, it recognizes stigmatized but essential workers on Bourbon Street as stakeholders in creating the experience of heritage tourism. This work opens questions about how such labor affects the lives of service workers who create those experiences, who, following Kirshenblatt-Gimblett (1998), export their work in the service industry. Intangible heritage, such as the French Quarter as a place for the historic and persistent market in intimate encounters, is a part of the tourist experience which is poorly understood. By focusing on practice, we can move past the visual and on to the complete sensorium of the tourist.

Performances that take place on Bourbon Street are an important component of re-creating what Frohlick (2007, 139) calls the "social sexual history" of the place in global tourism flows. In the case of New Orleans, sexual services and erotic experiences are a part of many tourists' imaginaries about the city, as they have been part of the reality of visiting New Orleans since before modern tourism. To make sense of the chapters that follow, we must begin by thinking of New Orleans as a port town receiving visitors rather than a tourist town receiving tourists. Dean MacCannell's seminal work on *The Tourist* (1976) defined modern tourism as a relief from the social alienation of the middle class, a search for the authentic, a search for the real in a spatially separated place of leisure. Although B-girls certainly engaged with such tourists by the 1990s, the historic record shows that the practice of B-drinking predates the modern post-World War II tourist era. B-girls in New Orleans served port workers, sailors, and traveling businessmen before modernity and such claims of alienation. Richard Campanella suggests that in the past as today:

> Sailors, travelers, visiting businessmen, and other transients, liberated by their anonymity and decoupled from the responsibilities and restraints of home . . . gravitate to opportunities for immediate gratification, in which alcohol plays a primary role. (Sex is a close second—and, not coincidentally, accounts for another historical reputation associated with this city. (2008, 295)

As I will show, B-girldom in New Orleans is associated with people engaged in mobility more generally long before it appears within the discourse specific to tourism development.

Engaging with a feminine companion who professionally solicits drinks predates the tourism boom, and it predates the creation of Bourbon Street in its contemporary form as a site for tourists. This book presents a case study of the historic and social processes involved in the creation of the stereotypes and reality of a highly gendered type of service-industry worker. The B-girls performance is an important component of intangible tourism in the French Quarter, where images of sexuality and imbibing in alcohol collide. The practice has long structured opportunities to collect money for off-stage encounters while masking the commodification of those intimate relations by controlling them using existing gendered behavior as they relate to liquor sales.

The story of B-girldom is about the structure of gendered interactions that are a part of the master narrative of Bourbon Street as a facet of New Orleans' image more broadly. Ethnographically, my participants told me about their experiences and how their choices (both short and long term) to work in the intimate tourist market as B-drinkers affected them. A unique category of service work, B-drinking successfully commoditizes performed sexualized encounters and fulfills the expectations of the individual tourist within the broader master narrative of Bourbon Street to "delve egoistically into the imagination while still interacting on some social and tangible level with a place" (Di Giovine 2009, 145).

TOURIST DESIRES AND THE RED-LIGHT EXPERIENCES

Many years spent in strip clubs on and around Bourbon Street lead me to agree with Vesey and Dimanche's statement that "locals do not visit Bourbon Street often, unless they work there, or are out for a night in the French Quarter with visitors from out of town who want to see the mythic adventure Bourbon Street has to offer" (2003, 66). Their notion of the Bourbon Street audience prioritizes the geographic (leaving home) and leisure. Gravari-Barbas and Graburn also explain tourism imaginaries as constructed "concerning places outside their [the tourists'] primary residence" (2012), following MacCannell (1976).

But locals do occasionally engage with the "mythic" and generally participate as tourists when they walk visiting guests down Bourbon Street. This phenomenon, the "lure of the local" (Lippard, 1998) removes the need for movement (and its expense), instead defining *engaging* as a tourist as "an alternative approach to material interaction, a perspectivally different lens" (Di Giovine 2009, 149) building on the theoretical idea of tourism as one of many frames through which one can experience a place. When locals do visit the Bourbon Street cluster, it is often in a "socially prescribed" (Di Giovine 2009, 153) way, such as a bachelor party, which includes a shift in perception to a celebratory frame. In early anthropological work, Valerie Smith (1989) also focused on the framing of the function of the experience (for example, to celebrate) in understanding tourism as being "for the purpose of experiencing a change."

It is mainly the desires of tourists on Bourbon Street that allowed for the success of B-girls who worked in the 1990s. The strategy of B-drinking can lead to conflict between local customers and employees, but catering to a short-term visitor's needs for erotic forms of emotional labor is a very different endeavor from engaging in performances of intimacy with persistent locals. Walter Little talks about the strategy of "tourism as performance" (2003, 529), recounting how Mayan women similarly build rapport using "Brechtian theatrical techniques" in which "the tourist becomes part of the play." Like B-drinkers, Little's informants live and work in a tourism site but are not in positions of formal power. Shifting their performance to cater to the particular tourist, they use humor and their position as bearers of local culture to minimize the visibility of their economic goals of making sales.

It is the temporality of the encounter that allows the women who work on Bourbon Street to maintain high incomes. At the same time, such performers depend on cooperative strategies that lead to the development of a strong sense of community among themselves. This contrasts with ethnography of striptease in other areas of the country where service workers in the striptease industry must deal with "regular" customers. As Julia Harrison points out, tourists often seek out intimate glances into their destinations, and in reminiscing about past trips focus on the "positive emotional memories and responses," what she refers to as the "touristic aesthetic" (2003, 93). As we will see, even tourists who joke about having been "ripped off by a B-girl" tend to remember their experiences on Bourbon Street in an overall positive light, as part of an intrinsically rewarding tourism experience.

Di Giovine (2009, 157) addresses the issue of nostalgia in his work on tourism, focusing like Graburn on tourism as a site in space and time where tourists experience rites of intensification, cyclically renewing the social order for those on tour. Di Giovine locates nostalgia in the idealized memories of tourists that contrast with current life. Although I will show evidence that Bourbon Street has been a site for such nostalgia for visitors, my focus is understanding the relationship between the construction of tourism and its effects on those who have come to work. Like tourists, service workers shift in and out of the tourist site, between normal life and the "differentiated way of perceiving one's surroundings" (Di Giovine 2009) that characterizes engaging in tourism.

WHAT B-GIRLS DO

In order to understand how exactly B-girls fit in with the tourism industry we have to understand what it is they do. B-drinkers carefully record the number of drinks that club patrons purchase for them; these are officially recorded by the waitress, bartender, or manager. At the end of a shift, the B-drinker (legally referred to for most of the twentieth century as a B-girl) receives a payment for a percentage of the cost of each drink the patron buys. Thus, a B-girl primarily works selling alcoholic beverages in a nightclub, for which she is paid a commission or salary by the club management.

Power dynamics between travelers and various stakeholders as hosts in tourist destinations are complicated when intimacy is desirable, but direct sex for cash exchanges is discouraged (Frohlick 2013, 157). As an alternative to direct purchase of intimacy, a B-girl strategy places purchases of drinks and the nightclub between the movement of cash and the gyrations and giggling the B-drinker performs as an income-generating strategy in a sexualized tourism space. In this way, direct exchange of the material payment for erotic engagement is avoided and instead organized around notions of gifted drinks, generosity, and having a good time. Gendered gifts of alcohol are the norm in American nightclub culture, and so alcohol sales as a mediating factor in exchanges of money for intimacy can be invisible to the tourist. B-drinking allows the service employee to translate the tourist's desire to have a good time into a wage, and it keeps the economic aspects of the performance of an intimate encounter veiled enough for a tourist to enjoy the suspension of disbelief necessary for lap dances, stage performances, movies, or any other sort of performance or storytelling.

IMAGE AND INCOME—DISJUNCTURE IN NEW ORLEANS: INTIMACY AND THE PERFORMANCE OF HERITAGE

Guided by participants in my research process, I recognize the importance of convention tourism for such businesses. Conventions, and other work-related tourism which brings large groups of professionals to the city, intersect with and support certain forms of sex work, such as that found in Bourbon Street strip clubs. In an article published by *ABC News* (Rood 2008) about expected spikes in the sex trade during political conventions, former sex worker and spokesperson for the sex-workers-rights group Call Off Your Old Tired Ethics (COYOTE) Carol Leigh said that political conventions, like any large convention, would mean a "boost for the sex industry." Other large crowds, such as festival tourists, may flood the streets in greater numbers but tend to spend less money and are therefore of little interest to my research participants.

Theoretically, I lean on Arjun Appadurai's (1996) idea that to understand the contemporary world it is important to recognize the different dimensions of global flows, originally suggested in his five "scapes." Each layer is fluid and changing and interacts with the other four kinds of flows in ways that often defy political boundaries, creating sites that are imagined contextually in a larger global network. Globalization is characterized by flows of people (ethnoscape), information through media (mediascape), technology (technoscape), fiscal capital (financescape), and political ideologies (ideoscape). In any geographic place, overlapping experiences suggest to the individual a mapped understanding of global realities, what Appadurai calls our "imagined world."

Critical to any study of work in New Orleans is the city's relatively weak economic position within a country that dominates the global financescape. As a poor city in a rich nation, New Orleans has for the past several decades been dominated by its tourism industry, which provides services and labor to people from all over the United States and indeed the world. Campanella (2014, 272) estimates that on a typical evening, 72 percent of Bourbon Street pedestrians are from out of state and 9 percent from outside the United States.

The statewide Louisiana Tax Free Shopping program for international visitors specifically caters to international visitors. Unlike duty-free shops in airports, the Louisiana program provides a refund of taxes on purchases of tangible items from more than 1,500 department stores, antique and art dealers, and manufacturer outlets to international visitors departing the country. In 1998, of the 25 countries with the most participation in the program, Honduras, Mexico, Germany, Brazil, England, and France were all countries spending $2 million or more (Dimanche 2003, 311). Dimanche (313) points out that one benefit of the program is that it allows the state to collect information on international visitors.

The growth in visitors can also be seen in the development of the convention industry, with an increase in the number of conferences from 172 in 1960 to 3,556 in 2000 (Gotham 2007, 15). In 2013, the city saw a 4.5 percent increase in visitors, with 9.28 million people visiting the city and spending 6.47 billion dollars (City of New Orleans Mayor's Office Press Release 2014).

Tourists are drawn by the ways that the city has been represented within the broader mediascape. This combination sets the stage for New Orleans tourists. Building on this theoretical notion of layering, other authors writing on tourism and the performance of sexuality have further "scaped" the world as anthropologists. *Sexscape* is anthropologist Denise Brennan's term for "a new kind of global sexual landscape . . . where the practices of sex work are linked to the broader forces of the global economy" (2004: 15–16). New Orleans' French Quarter has a known role in the American sexscape, and both workers and customers come from around the country (and sometimes the world) to take part. This node in the international economy overlaps with broad areas of the New Orleans *touristscape*, defined by Erica Williams (2013) as those areas of the city where "the tourism industry is a central focal point . . . (along with) the many people who occupy and move through those spaces," (see also Desmon, 1999, 14, and Metro-Roland 2011, 38, for earlier discussions of the touristscape).[7] Tourism revolves around attractions like Bourbon Street, and the employees and contracted workers of strip clubs are part of the unique performance of Southern heritage found in the contemporary French Quarter of New Orleans. New Orleans' heritage tourism landscape and the economy of desire in America are intertwined to the convenience of the male business traveler and to the enrichment of the international level service industry stakeholders of the region.

I am mainly concerned in this text with understanding one type of host (Smith 1977) among a myriad of service providers; B-drinkers and striptease dancers work alongside waitstaff, taxi drivers, restauranteurs, and hotel employees who all serve the city's guests. A holistic understanding of the situation of Bourbon Street performers requires consideration of the perspectives of a myriad of stakeholders, such as city and state officials, museographers, and historians who contributed to the construction and control of the tourist economy. A historic look is required to understand the processes that have contributed to modern Bourbon Street tourism performances.

Living in the French Quarter quickly reveals that the cosmopolitan nature of the neighborhood is more complex than a simple "local" identity, as the neighborhood continues to attract newcomers from around the country and the world. It is an attractive and busy community, thriving and blended. Still, host and guest as a binary remains a useful heuristic device for understanding the perspective of the tourist, who tends to believe the binary to be a fact. It also reflects the undeniable division between those who perform service work as their livelihood and those who are temporary visitors and customers of the service industry as a part of their holiday calendar.

Central to the prosperity of B-drinkers is the guests' desire to have meaningful, close, and intimate relationships with local people, the regional hosts. These hosts are those who live near the heritage tourism site (in the case of Bourbon Street, sometimes in the farther suburbs of the city) and earn their incomes by striving to create an "authentic" or enjoyable experience that conforms to their expectations of the destination for those who visit from afar. For the hosts, the attraction of heritage tourism is economic; the past contributes to the creations of new ways to make a living by selling goods and services. In addition to goods and services, the hosts also offer to tourists the less tangible experiences of being in historic or historic *feeling* spaces, an experience of a culturally constructed ambiance or an imagined re-creation of the past. This feeling, which heritage sites hope to induce, has been discussed as "nostalgia," an important component of the tourist experience (Frow 1991, Graburn 1995, MacCannell 1976).

An important component of the creation of the French Quarter is the emotional labor done to invoke feelings, like nostalgia, that are sought by the tourist. This is offered to guests as friendly interactions with personnel who somehow "fit" into the specific historic and cultural context that the site strives to re-create through carefully constructed performances of their gender, race, or class. Heritage tourism, and tourism work broadly, includes a tremendous amount of emotional labor (Hochschild 1983). Tourism professionals have paid increasing attention to occupations that include such emotional labor, such as waitstaff in restaurants, hotel reception desk workers, and a myriad of other jobs, with an eye toward managing its deleterious effects. This kind of work is related to emotional exhaustion (Chu et. al, 2012) and poor job satisfaction (Gursoy et. al, 2011). Emotional labor demands communications that are controlled to express or suppress emotions as part of the job, and in such encounters within the service industry, customers' emotional states were found to be directly affected by the "deep acting" (Hennig-Thurau et al. 2006) of service providers perceived by customers as reflecting "authenticity of employee's emotional labor display." The tourist desires to be served physically *and* emotionally, expecting many kinds of workers to perform concern and pleasure while at work within the view of the tourist. B-drinking is an extension of how this market need is fulfilled and how working hosts are compensated for their labor without a disruption of the performance.

Heritage tourists want to experience a site for its historic or cultural tradition, unique or significant architecture, civic planning, landscape design, food and drink, music, art, ambiance, or cultural performances. Such sites have some unique historic or cultural qualities that visitors find attractive. An important part of that experience for heritage tourists is often the local people. Enjoyable interactions with both professional hosts and other "real locals" can help to make a visitor's experience of the place feel like a more authentic one. This study of the strategies taken by women on Bourbon Street that allow for the commercialization of the intimate illuminates some of the ways that tourism studies and anthropology intersect.

I focus on employees of the so-called strip clubs, nightclubs, or, more recently, gentlemen's clubs that dominated Bourbon Street at the end of the twentieth century. I introduce to the existing body of literature on erotic labor and performance a historically long-standing staple of New Orleans sexual economy: B-drinking. B-drinking is one way that visitors can interact with tourism professionals, who perform the emotional labor of being enthusiastic and amiable companions, mainly for the male tourist.

Anthropologists today are fashioning an increasingly nuanced understanding of the complex intersection of tourism and sex. Current ethnographies of how sexuality is marketed in the global tourism economy go far beyond a simplistic discussion of sex as labor to acknowledging the diversity of practices in specific cultural and historical contexts. Much of this literature relies on a multiplicity of ethnographic descriptions, complicating the picture by contextualizing specific forms of erotic labor within particular understandings of ethnicity, gender, and indeed the cultural meaning of specific sexual acts (see for example Brennan 2004, Frohlick 2013, Kulick 1998, Kempadoo and Doezema 1998, Simoni 2015, Wardlow 1998, Williams 2013).

Adding the New Orleans example can broaden our understanding of how particular forms of sex work evolve in particular places and in relation to tourism. Like any industry, the development of contemporary striptease bars in America differs by regional histories. The local history of place must serve as a starting point for understanding the contemporary political economy of New Orleans' French Quarter.

In a contrasting contemporary ethnography of strip clubs, Susan Dewey (2011) locates her own observations in a city in the American "rust belt." Dewey's *Neon Wasteland* is an ethnography of the American strip club but positioned in a part of the United States remarkable not for tourism but because of its recent industrial boom and consequent bust. Cities like Detroit, Cincinnati, and Cleveland have gone through a process of first being centers of manufacturing and now sites of urban decay as industries have moved elsewhere. It is in this place—embedded in Dewey's title—that the strip club she describes operates. This is very different from New Orleans, a port city that has survived the boom and bust of oil by catering to visitors charmed by the city's colorful past, Caribbean flair, and reputation for the arts, leisure, and vice. After all, to cite the well-known quip often attributed to Tennessee Williams, "America has only three great cities: New York, San Francisco, and New Orleans. Everywhere else is Cleveland."

The expanding ethnographic literature on striptease in the United States since 1990 reflects the growing market for sex-related goods and services broadly. According to Eric Schlosser's essay *An Empire of the Obscene,* "Americans now spend more money at strip clubs than at Broadway theaters, regional and nonprofit theaters, and symphony orchestra performances—combined" (2003, 113–4). Although Schlosser's journalistic style of writing may have led to an overestimation, anthropologist Judith Hanna noted in 1998 in "Undressing the First Amendment and Corseting the Striptease

Dancer" that at that time there were around three thousand strip clubs in America.

Other ethnographers of striptease who have used a participant observation approach have taken different perspectives on social interactions within nightclubs. R. Danielle Egan (2006) specifically focuses on the relationship between dancers and their regular customers. Katherine Frank (2002) also addresses the relationships between dancers and their customers, arguing that the clubs are sites for public, voyeuristic interactions for her male informants. Frank's interviews focus on men as customers, selected for her sample because they visit such clubs "often enough to consider this a significant personal practice" (2002, xxiv). Rather than focusing on such increasing "commodification of spectacle and experience" (xxv) from the perspective of the motivations of customers, my ethnographic point of focus is the perceptions of the women of themselves and their own relationships. Additionally, rather than the interactions between customer and service worker, I am more interested in the interactions among service workers as a community and their experiences of the positive and negative aspects of their labor.

Finally, on a gloomy note B-girls provide us with an unfortunate case study of structural violence in the United States. This example illuminates the powerful tendency toward structural violence against women, both in the daily lives of my participants and in the use of laws against B-drinkers. Such laws have historically been stubbornly gender specific. Some of the structural issues faced by strip-club employees, such as labor rights violations, are common to the service industry of the region. Much of the situational violence that is a part of the B-girls' life is familiar to many low-income French Quarter employees, including tour guides, massage and other spa providers, waitstaff, kitchen staff, and many others who provide direct services to tourists. Given the weak economy of New Orleans and Louisiana more broadly, there is often a significant income gap between those who provide service, and those who receive them. This is highlighted by large groups of professionals, including anthropologists like myself, who attend the elaborate conferences hosted by the city and support the economy of the French Quarter. Here, through a detailed look at one specific sort of service worker, I hope to enrich our understanding of tourism work generally to better understand the situation and experiences of those who provide other sorts of personal/intimate tourism services.

Within the context of New Orleans, the construction of the B-drinker in the public imagination has played a critical role in the structure and control of sex work on Bourbon Street. This feminine workforce is in a position of power in the functioning of the space, which as a heritage tourism site depends on their performance of promiscuity and commodified sexuality. Almost every tourist to the city at least takes a walk up the famed Bourbon Street, a space where vice and nudity are expected.

The women who dance in windows, behind screens, or whose photos are posted outside nightclubs are (almost entirely) a contract workforce. Classified as independent contractors and forming tightly knit social groups, the semi-nude dancers of Bourbon Street far outnumber the owners and management whose task it is to control the workplace. This leads to employees holding a great deal of power in decisions about what the subculture within the club will accept; in the clubs under observation, female contract dancers set (and sometimes enforced) internal and often tacit rules on wide-ranging topics such as acceptable forms of sex work, the pricing of specific sexual acts, the use of drugs while inside the club, and what percentage of a customer's money would make its way to the cash register. Their agency must be understood, however, within the limits of the illegality of their work.

Perhaps counterintuitively, the group power as stakeholders that cooperating strippers have to control their own work environment is countered by the requirement to B-drink. Requirements that employees solicit and accept drinks (alcoholic or otherwise) shift the dancers from contractual talent to criminals, vulnerable to increased stigma and the sporadic threat of arrests. I believe the illegality of the practice reinforces the male/management centered rules, but it also tightens the feelings of group cooperation among women working together. This creates an ongoing struggle over use of space as it functions, and seems to have functioned, for many decades. Laws against B-drinking are used to counteract the agency of women who question working conditions and income distribution in the industry rather than to actually limit a practice (required alcohol consumption as work) that could be perceived as harmful to employees.

BOURBON STREET SCHOLARSHIP

A search of the literature on Bourbon Street, New Orleans from the perspective of a scholar would suggest that the area is mostly populated by musicians. Anyone who has worked on Bourbon Street, or in the surrounding neighborhoods of the French Quarter, knows that musicians are only a small fraction of those who earn their livings in this neighborhood. Increasingly, researchers around the United States have turned their attention to striptease dancers (or strippers, exotic dancers, burlesque dancers, etc.), legitimizing the research of this gendered occupation. Although there is less written on contemporary sex work in New Orleans than in the western United States, the topic of sex work generally is strongly represented in contemporary feminist scholarship.

The academic literature might also lead one to believe that B-girls are a thing of the past, as they receive mention mainly in some historic articles on regions around the United States. A search on B-drinking in New Orleans will reveal infrequent arrests, suggesting that such behavior is rare. However, there are many mentions of B-drinkers in literary narratives of the city. My observations in New Orleans between 1992 and 2001 revealed that the practice legally termed "B-drinking" was an important source of income for the women who work as dancers, waitresses, and bartenders in many clubs across the touristic areas of New Orleans. Ostensibly contracted legally as dancers, more than half of the income of women interviewed was earned through the grey market of B-drinking.

Understanding the economics and social conflicts faced by the B-drinker is one example that contributes to our understanding of tourism-related sex work in New Orleans, as well as along the Gulf Coast region more broadly. Archival documents demonstrate that B-drinking, though not always by that name, has long been a common practice in New Orleans. My study of the historical evolution of B-girl heritage in New Orleans across a broad spectrum of sources suggests that the "B-girl" as a discursive construction has long played an important role in the French Quarter. Historically, B-girls appear as potential problems in discussions concerning the development of local/visitor relations; this is particularly true at moments of moral vulnerability, such as during war, when visitors marked as particularly susceptible to B-girls include soldiers.

Fictive narratives about the French Quarter commonly include B-girls. Plays, novels, and society columns all contribute to the expectation of B-girldom as a part of the French Quarter. Even for those who never step into a strip club, Bourbon Street remains a central component of the tourists' cognitive map of the city. The industry workers who populate that space are critical components of the French Quarter as a touristscape, whose labor (re)creates French Quarter heritage.

B-drinkers inhabit an ambiguous structural position in the economics and performance/construction of heritage tourism in the French Quarter of New Orleans. A necessary evil for the stakeholders in New Orleans tourism, B-girls are, in life as in literature, both embraced and spurned. B-girls inhabit a sort of double-bind. They are recognized as an entertaining component of the authentic French Quarter experience, "quaint creatures" (Hopper 1948, 24, March 26), who seem to have always embodied the nostalgia for a recreated past that drives local heritage tourism. They are also criminals, subject to arrest as well as social stigma, and they are legitimized targets for violence because of their known engagement in the illegal and polluting occupation of B-drinking.

This text increases our understanding of exotic dance by widening the ethnographic lens, looking past striptease itself and to the economic activities of such workers during the much-longer periods of time they spend between sets. The structure of B-drinking makes otherwise legal jobs, such as exotic dance, illegal. I found labor conditions in clubs to be poor, both because workers feared reprisal for making reports and because they were aware that labor laws did not apply to them as contract workers. In clubs where pay was organized around B-drinking, dancers' incomes increased beyond their stage tips. At the same time, they became marginalized tourism employees due to the requirement that they break the law through B-drinking.

Women's personal experiences as "B-drinkers" must be understood in a local historical and cultural context. The discursive B-girl, a prototype who exists in newspapers, legal documents, and fictional accounts of the city, affects the self-perceptions and reality of women actually employed in occupations that require B-drinking. Positioned as both celebrated and criminalized characters, a strong sense of community develops among women who share this locus in the touristscape. The fact that support networks of women formerly employed as B-drinkers remained in place 15 to 20 years later is a testament to the intensity of relations built by women who worked together on Bourbon Street. Dressing-room intimacy results in a feeling of community strong enough to be maintained over decades.

WOMEN'S STUDIES: OF FEMINISM, SEX WORK, AND FEMINIST SEX WORKERS

Among feminists, there exists a deep division between those who argue that sex work is inherently damaging and those who argue instead that it is the conditions that sex workers must tolerate that are damaging. Differently motivated feminists have advocated for women's rights working within the industry, as well as broadly for the protection of women from the perceived harms of sex work through a strong abolitionist movement that aims at the eradication of sex work. This extensive literature, however, does not currently contain any documentation of B-girls in the New Orleans French Quarter, leaving this an invisible form of labor. This book, I hope, fills that gap.

The tradition began in 1871 with Josephine Grey Butler's "A Letter to My Countrywomen in England," which viewed sex workers as victims and laws meant to control their activities as harmful to women as a social group more broadly. Feminists approaching sex work, weaving left and right through theories, have arrived at a stark division. There is a near binary between the writings of contemporary advocates who prioritize the rights of sex workers as laborers and the abolitionist stance most famously articulated by Andrea Dworkin (1981) and legal scholar Catharine MacKinnon (1989) in their argument against pornography.

The debate is ongoing among feminists, with some scholars completely rejecting the notion of erotic labor as work and other's demanding that sex work be decriminalized so that women who choose sex work can be better protected. Taking a perspective of identity politics, some sex workers have articulated to me that they feel abolitionists are paternalistic and demeaning in their assumptions.

Two politicized opinions have developed when it comes to the way the legal system, social services, and the medical community do and have historically conceptualized sex workers. Carol Leigh (1997) describes this division as based in the "internalized oppression" experienced by women, which leads to horizontal hostility among them based on patriarchal categories that silence and censor some women (1997, 228). Leigh points out *many* women "trade sexuality for survival, or some social advantage" (229), whether as wives or sex workers. From her perspective, this divides women into types within a patriarchal folk taxonomy, "based on the contracts we made with men for the purposes of our survival" (1997, 229). Leigh (230) suggests that even sex workers of different types are pushed into such horizontal hostility and offers the term "sex work" as a "feminist contribution to the language," meant to "unite(s) women in the industry—prostitutes, porn actresses, and dancers—who are enjoined by both legal and social needs to disavow common ground with women in other facets of the business" (230).

Today, the two camps have a great deal of difficulty communicating with each other, although both are rooted in feminism and look to protect women from harm. Considering a similarly stigmatized population, the developing genre of self-published autobiographies (which have overlapped with my research when looking retrospectively at the 1990s) reveal a binary positioning of the heroin addict as either diseased or a criminal. Women who are sex workers are also portrayed in such a binary. They are either vulnerable and in need of rescue or despicable gender traitors.

After years of interactions with various kinds of sex workers, I am uncomfortable with such a simple binary. As Jack Katz (1997) notes, social distancing occurs when a group (like strippers, B-girls, or prostitutes) are viewed by society as posing a risk, and social scientists justify their interest in a group or process by one of two characterizations. The first is a portrait that demonstrates that the anxieties about the group are unfounded, thereby normalizing their behaviors because "groups thought to be deviant actually are serving conventional motives" (1997, 395). The second way is through "bohemian" writings, which report the exotic nature of the subject and present the group as more deviant than already imagined. While those engaged in harm prevention discourse tend toward the first, conversations driven by the second motive are the focus of abolitionists. As an ethnographer, I try to avoid either portrait, because I found that Bourbon Street is a poor model for either such extreme perspective.

Both camps of feminists have similar goals. Both hope to take a second look at the laws as they apply to people who earn money by performing erotic labor. Both hope to create better lives for women in their communities by minimizing the harm that is experienced by women involved in or otherwise affected by the sex industry. Where they differ is in what steps they believe must be taken to reduce harm and often in opinions on who is ultimately within the circle of those affected by the many aspects of the sex trade as it exists in their communities.

The divide in feminist approaches revolves around the issue of choice. Some theorists take at face value the personal agency reported by many sex workers in choosing their occupation and hope to support those workers through a rights-seeking approach. Opposing are those theorists who see sex work as inherently disempowering, and who view sex workers who claim agency as having a sense of false consciousness that prevents them from realizing they are victims. This has led to an oversimplification of the complex issue of sex work, choice, and gender in American society within the academy.

Abolitionists are those who would seek the elimination of the commodification of sex in their communities, viewing all such work as essentially damaging to both the sex worker herself and women in the sex worker's community. The abolitionist agenda suggests that the acceptance of sex as a commodity hurts women in two ways. The woman who receives an income is assumed to be victimized by sexual interactions that are commoditized. It does not matter, from this perspective, whether the sex worker herself *feels* she is a victim, or damaged. Rather, if she does not, she is imagined to suffer from a sort of secondary victimhood of false consciousness. Her false consciousness prevents her from understanding her position in the broader economy because of the structural power of her abusers.

Abolitionists are often strongly influenced by the work of feminists like Andrea Dworkin, whose book *Pornography: Men Possessing Women* (1981) accuses the pornography industry of doubly abusing women. First, female porn stars are abused, according to Dworkin, in the production of the film. Second, and in an ongoing fashion, pornography is believed to entice men who view it to similarly abuse women themselves.

Closely associated with Dworkin is lawyer Catherine MacKinnon (1993), famous for her development of "dominance theory" in thinking about the relationship of women to the law. Whereas her precursors argued about what patterns of law, formal or substantive equity goals, would ultimately create better lives for women, MacKinnon believed that understanding American law required a deeper look at the relationships of power and sexuality in the United States. Pornography, and with it other forms of sex work, was viewed by this theoretical approach as damaging not only to the women directly touched by it. Pornography is, more broadly, a perpetuation of a systematic oppression of women through culturally proscribed sexual roles. Thus, for those who might have never knowingly met a sex worker, dominance theory proposes that it is still directly affecting their lives.

> To express eroticism is to engage in eroticism, meaning to perform a sex act. To say it is to do it, and to do it is to say it. It is also to do the harm of it and to exacerbate harms surrounding it. (MacKinnon 1993:33).

Catharine MacKinnon's groundbreaking work on dominance theory is a well-argued treatise which insists that the law recognize the intersection of power and sexuality in the workplace. This argument is well applied to cases of sexual harassment at work where men tend to be in supervisory positions over younger women and where emotional labor can be tacitly required and abused. Thus, the law ought to protect working women from those who are economically stronger and who would quietly coerce them to accept as natural others' misogynistic views of their sexuality. By this argument, the way to do this is through control and illumination of the intrinsic abuse of sex work and other forms of gendered labor.

The problem here is it does not hear the voices of women who are sex workers. There are multiple narratives told through autoethnography, journalism, and cinematography created by women who work or have worked within the sex industry and feel they did so by choice. From this perspective, it is paternalistic to suggest they cannot be agents in their choice to engage in occupations that include some form of sexual performance. In "Forced to Choose," Jo Doezema (1998:35) suggests that the discourse of choice has been seemingly constructed to define and fight potential trafficking of sex workers, but fails to untangle the complexity of defining victimhood in a system where all sex workers must be either trafficked victims or sluts. Anti-trafficking efforts and the abolitionist perspective have contributed to a false dichotomy between "innocent" victims who are forced into sex work and "guilty" sex workers who choose such labor. Doezema also points out that this dichotomy is an oversimplification of women's actual experiences dealing with different sorts of "forces," such as the force of economic need that drives people to many less-than-optimal kinds of work.

As anthropologists begin to untangle the multiplicities of hard and soft forms of power, it is certain that pornography like any other film contributes to gendered scripts of sexuality for both men and women. However, the hard power of economics forces many working-class Americans to engage in daily labor that they may not otherwise choose. Few people bother to ask their fast-food cashier whether he or she feels coerced to do their work, or if they had to relocate to find work opportunities. Across most of the American labor market, it is assumed the individual employee has made an informed choice to engage in work for hire. Why is this not so for sex workers? Why is hard economic power interpreted differently in this case?

Kamala Kempadoo and Jo Doezema's (1998) edited volume includes chapters in which sex workers from Cuba, Japan, Ghana, Thailand, Ecuador, and Mexico situate their own experiences of sex work in a similar framework of agency. The volume includes stories of regionally specific attempts to survive or advance through engaging in sex work, as well as sex workers' attempts to organize and further empower themselves as groups of legitimate workers in various national contexts. In another excellent volume, *Sex Work: Writings by Women in the Sex Industry*, first published in 1987, editors Delacoste and Alexander (1998, 2nd edition) provide space for sex workers and sex-work activists to tell their own stories. The volume includes chapters written by street-based prostitutes, massage parlor workers, erotic dancers, and women who do live sex shows. All fall under the umbrella of sex work.

Sex-worker's rights activists, in contrast to abolitionist theorists, tend to reject the notion of false consciousness in sex work. They include among their ranks famous sex workers turned activists whose personal experiences lead them to reject an abolitionist approach. Women like Carol Leigh and Annie Sprinkle, speaking from their perspectives as individuals who have earned their incomes through sexual forms of labor, provide a sex-positive foil to the voices of abolitionists. They argue for a harm-reduction approach that would protect women who engage in sex work from the various forms of abuse to which they are particularly vulnerable, including from law enforcement.

Since the emergence of feminist jurisprudence in the 1960s, we have begun to examine the implications of seemingly neutral laws and practices as they affect employment, family law, reproductive rights, and many other issues in gendered ways. Here, similarly, I hope to interrogate the relevant laws, the evolution of those laws, and their application among women who engage in B-drinking as part of their work. Economically, B-girls are empowered, and as we shall see they are particularly empowered within the space of the nightclubs and Bourbon Street itself. Historically, the self-regulation of sex workers has been diminished through the manipulation of laws that indirectly address their economic power and/or limit their power to organize as collective associations of workers. I hypothesize that the current functions of laws (city and state) against B-drinking were used in the late twentieth century in a similar way. Though seemingly rarely enforced today, this antiquated legislation can be used at any time to arrest otherwise legally employed or contracted entertainers who are required to illegally drink on commission to keep their jobs. In its current state, the law serves as an important crutch for the small number of (mostly male) bar owners, managers, and beat police officers who maintain power over an overwhelmingly female labor force.

Sex work is a broadly defined category, used here as a term of inclusion to shift the discourse about women who engage in stigmatized forms of labor away from an identity label and toward recognition as *workers* whose labor is integrated into a broader economy. Striptease as a form of sex work has been increasingly stigmatized as a part of *America's War on Sex* (Klein

2012), which has included attempts to close or limit striptease by proponents of the "Christian Right" (Hanna 2012). B-drinking falls within the broader category of sex work, a term coined by Carol Leigh in 1980 and popularized in 1987 (Delacoste and Alexander 1998) to encompass the work of prostitutes, exotic dancers, escorts, and other laborers. My intention in using the term is to increase our understanding that labor does not equal personal identity and to highlight the feelings of many in the industry that such work is not self-exploitative.

I'm grateful to a reviewer who pointed out the ambiguity of the term. Sex work is, in fact, ambiguous in the New Orleans French Quarter. Most B-girls in the 1990s also engaged in striptease, an occupation within the realm of sex work as defined by Leigh (1997). It is work that is broadly understood as meant to sexually arouse. Much of the conversation between a B-girl and her clients is sexually explicit, or at least suggestive in nature. Depending on the sexual act performed, the same woman can slide between legal categories of misdemeanor prostitution and felony crimes against nature—that is, colloquial definitions of "prostitute" do not entirely overlap with legal definitions of the term. None of my informants in the 1990s could explain the legal categories, and in fact some of the acts that were considered less offensive and more common (oral sex) held stronger penalties in the state of Louisiana than acts that were never observed within the club but would have been considered more serious within the community under investigation (vaginal sex).

The term "sex worker" is inclusive of those who engage in many forms of sexualized labor, whether the performance includes a penetrative act or no physical contact at all. It is further useful to me in its focus on labor and labor rights rather than legal classifications or identity. Aside from a category of person rather than of work, "prostitute" is a legal term with very specific definitions that vary by region. Sex work can include work with no physical contact, such as telephone sex, as well as work that might include various acts of sex for sale. In Louisiana, behaviors colloquially lumped as all "prostitution" are in fact legally separated between prostitution (a misdemeanor) and acts "against nature"—oral or anal penetrative contact (a felony). B-drinkers share with other sex workers a wide range of emotional labor that creates the phenomenological experience of a sexual encounter or simulation of one for the buyer, *however he might define it*. Confronted directly with the question of whether or not B-drinking was sex work, my friend and interview participant Jade laughed. Unlike other performers, such as a film actress who engages in sex as a *part* of her performance, Jade pointed out that as a B-drinker *all* of her performances were erotic in nature.

In understanding my informants' choices, rather than agency being a simple on/off switch, I heard workers convey a message of complex decision making in which individual acts of sex work were negotiated with considerations of risk, potential profit, seasonality, and often with a deep understanding of gender stratification in the United States. Rather than a simple matter of choice or cultural coercion, the decision of whether and how to engage in sex work is viewed by the workers themselves as a complex process, fraught by variables both social and individual.

I include B-drinkers in the cultural domain of "sex worker" based on my long-term participant observation in Bourbon Street nightclubs. B-drinkers are women who engage in the sale of services that are erotic in nature, with a professional understanding of sexuality as a culturally constructed domain. Through the voices of my informants as workers, the messy issues of incompatible theoretical models disappeared and were replaced by pragmatic decision making concerning morality and economics in a gendered world.

IS STRIPTEASE AN ADVERSE SECONDARY EFFECT OF B-DRINKING?

In 1986, in the case of *Renton vs. Playtime Theatres, Inc.*, the Supreme Court of the United States wrote that it was not the content of the films shown but the "secondary effects of such theaters in the surrounding community" that were objectionable. The doctrine of secondary effects was created in a social and political milieu in which the regulation of sexually oriented business to limited geographic zones was argued to protect neighborhood residents from crimes such as theft or losses in their property value due to detrimental activities (including prostitution and drug use) in their neighborhoods. French Quarter residents, however, generally benefit from tourism incomes. Questions concerning the dangers of public nudity that take root in other regions fall upon deaf ears in New Orleans. In the 1990s and today, the neighbors of such clubs do not view stripping as having adverse secondary effects in their neighborhoods. Many of them earn incomes in direct and indirect ways from both tourists and the employees of such clubs.

B-drinking is not an "adverse secondary effect"[8] of the Bourbon Street cluster of clubs where the strategy was found during the research for this book. B-drinking predates the current bars and was a known form of available, intimate labor before and throughout the development of these locations during the twentieth century. It was not until well after the post-World War II era of Bourbon Street's glory that the doctrine of secondary effects was even introduced into legal discourse in attempts to regulate strip clubs. Across the United States, the First Amendment protections which striptease had gained as it was increasingly recognized as a form of symbolic communication were countered by new ideas introduced into the law about the need to protect neighborhoods from hypothesized negative effects that were secondary to sexually oriented businesses. In considering the timeline of Bourbon Street, secondary effects as a national concern did not enter the discourse on strip clubs until they were well established and arguably already in decline on Bourbon Street.

LANGUAGE, CONSTRUCTION OF A TYPE, AND TACITLY DENYING WORKERS' AGENCY

Jon Donlon and Jerome Agrusa's (2003, 119) report on the "naughty" clubs of the Quarter wrote of nightclub management, "They are in business to make widgets to sell to widget buyers; these widgets wear tassels." Such tongue-in-cheek turns of phrase are entertaining, but they thinly veil the assumption that dancers are sold objects. Merriam-Webster defines it as "an unnamed article considered for purposes of hypothetical example," or simply "gadget." Leigh (1997, 230), in discussing prostitution politics, notes that such language denies sex worker academics the ability to be heard. She asks of such objectifying language, "How could I sit amid other women as a political equal when I was being objectified like that . . . obscuring my role as an actor and agent . . . ?"

Misogynistic, objectifying language aside, Donlon and Agrusa make the important point that topless nightclubs are in fact a part of the overall pull matrix of the city. Whether such clubs are explicit destinations, they remain "implicit elements in a larger 'aroma of exoticism' wafting from the tradition of naughtiness suggested by Bourbon Street, Storyville, and related historic material" (2003:125). Even for those who never enter a strip club, almost every tourist takes the required stroll up Bourbon Street; strippers and B-drinkers attract tourists, even the missionaries who come to the city by the church busload to perform their own morality shows on Bourbon Street. The Christian Right simply contributes to the cacophony of performances found on Bourbon Street during crowded tourism seasons, bearing bullhorns and life-sized wooden crosses. My favorite cross included an impressive digital display of applicable Biblical quotes on the cross bar.

In the academic literature I find my subjects as objects. This example of writing that identifies female service workers as manufactured "widgets" dehumanizes them by labeling them as literal objects and further contributes to their marginalization. Language and labels are important to understanding what is happening in the French Quarter and to women who work there. In their article, aimed at a tourism studies audience, Donlon and Agrusa suggest that Bourbon Street is an example of a wider phenomenon in which carefully controlled "club clusters" can be part of multiple attractions to any tourist site. Bourbon Street is, however, not an example readily applied to other cities. Such objectifying language denies the real power these independent performers have as agents to support or shut down a business—by voting with their feet, and suddenly moving to another club, tassels, sequins, and all.

B-girls demonstrate that an understanding of sex work and what it involves must be contextualized in the specific region, within its own history of the political creation and consequences of the organization of such work. Far from a typical American "cluster" of nightclubs, the strip clubs of Bourbon Street have developed over decades, next door to restaurants and hotels all of which share deep historical roots in the French Quarter as part of the wider city. The "strippers, with their particolored sequined Brazilian slings and bulging saline bags," and the woman who "slithers across the stage's worn linoleum on her belly like a reptile" (Donlon and Agrusa's 2003:122) are neither objects nor animals, but rather tourism workers definitional of the New Orleans touristscape.

Feminist authors have noted that sexual services are commonly part of the gendered tourism industry broadly (Enloe 1990). Like Erica Williams' study of Bahia, Brazil (2013), for New Orleans visitors the touristscape is characterized in part by the "possibilities for sex." It is upon exactly such possibilities that B-drinkers capitalize. Donlon and Argrusa suggest that "close scrutiny by local authorities helps guarantee that investors/owners are reluctant to risk the sustained income and durable return by illegally allowing real sex in these faux sex entertainment environments" (121). This statement, of course, assumes that New Orleanians agree with these writers' view of what constitutes a "controlled" cluster.

WHO DARES RISK INVERSIONS AND POWER IN AMERICA? ON CHOICES, AGE, AND AGENCY

The advantage of doing research within American culture is that it is less "exotic," perhaps closer to home and our own understandings of the world than much other ethnography. One is a bit more familiar with the history and culture and can address a problem more closely. The veil of the "exotic," which in the past has allowed ethnologists to suggest explanations for behavior that do not assume similar patterns of logic, suggesting sometimes fantastic differences in the physiology and behavior of subjects and observer, disintegrates when we realize that we are analyzing the behavior of our next-door-neighbors.

The sex work that occurs in the French Quarter, as I have observed it, is the result of logical decisions made by people with free will, not simply a process that is done *to* its "victims." Hegemony, deep-seated power structures that are nearly invisible, such as are suggested by MacKinnon, obviously exists. People often know this, speculate about it, challenge it, resist it, or choose to work within its confines. The women who I write about here made decisions, variously informed, on whether to provide sexual services for cash incomes. These are complex decisions requiring a great deal of thought about the cost and benefit of the work. This is often the case in work that is a part of the informal economy.

Sex workers, like drug dealers studied by Philippe Bourgois (1996), face the predicament of finding better employment for themselves in the informal or shadow economy than what is available in the formal economy. Bourgois points out that without the informal (untaxed) sector, it would be impossible for the neighborhood he studied in East Harlem to economically exist and appear as it does; official statistics suggested his informants "should have been homeless, starving, and dressed in rags . . . according to official economic measures, well over half the population . . . should not be able to meet their subsistence needs" (1996, 2). This is also true in New Orleans, where work like B-drinking is available to women who for a variety of reasons need a steady income—mothers, travel-lovers, heroin addicts—but who cannot meet the requirements of a formal-economy job that would allow them to afford a reasonable quality of life.

There are many requirements to holding most formal-economy jobs in our culture. A key requirement is dependability. One must be able to work on a set schedule. In the world of sex workers, from lap dancer to prostitute, it is very easy to live an unscheduled lifestyle. This was true for women engaged as dancers and B-girls in the Bourbon Street nightclubs. Because the sex worker as a contractor "owns her own business," many of the industries' formal-economy branches cannot easily enforce a set schedule. For my participant Barbie, a legally blind, single mother at age 22, being a dancer in the 1990s allowed her to work around her childcare obligations.

Women in my sample earned substantially more than the minimum wage. For a woman without education, or even perhaps without any legal identification, nightclub work allowed her to earn a high enough wage to work only a few nights a week. B-drinking increased a dancer's income enough to afford a small apartment and provide exemplary motherhood by being "at home" most days, following a script of the conservative American family. Dewey (2011) reports on dancers in the "rust belt" who take turns babysitting in the evening so that they can likewise take turns working as dancers. This is a common reciprocal exchange on Bourbon Street as well. No one wants to hear that the sex industry in America feeds lower-class children, but it does.

None of my interview participants reported feeling coerced in any way to do their work. Women across my sample who were compensated to drink only, drink and dance, drink and sell sexual intercourse after hours—none ever indicated to me that they felt they were forced to do such work. In most cases, women I interviewed reported feeling increased rather than decreased agency overall while employed in nightclubs. They chose their work based on incomes and job flexibility, which gave them a greater ability to make controlled decisions about many *other* aspects of their lives. None reported a lack of agency, all felt strongly that they chose their work, and many interviewed in 2012–2014 had feelings of nostalgia when remembering their time spent working the nightclubs in the 1990s.

Two of the women I knew well in the 1990s told me during interviews in 2013 and 2014 they began working in the clubs as minors, under the age of eighteen, and could therefore have potentially been defined as victims of domestic minor trafficking. The word "trafficked" objectifies the person, grammatically assigns their movement to some agent outside themselves; these girls report having trafficked themselves. I do not doubt the existence of people trafficked for their labor, sexual and otherwise. I just say with certainty that no woman with whom I came in contact in more than 20 years of conversations with nightclub B-girls ever fit the description of having been forced by another person to move for work or forced to engage in sexual or any other kind of labor. The closest exceptions, perhaps, were the inept attempts at coercion to go to the club by drug-addled boyfriends. Most often, such young women reported happily leaving such boyfriends at home for the peace and companionship of the dressing room.

How can such women be "trafficked" at all, then? In law, we generally deny that people under the age of 18 can decide to leave home without their parents' consent. In both cases here, the girls were probably 17 when they began working in nightclubs, and in both cases told harrowing stories of having fled homes where abuse occurred. Underage runaways are fugitives, who before age 18 must learn to economically support themselves while avoiding the paper trail of legally existing anywhere. Such young women face some of the same problems as undocumented immigrant workers. They want to work, they must work to support themselves, but they are severely limited in their opportunities by the real risk that they could be discovered and arrested. It should not be surprising that age is a taboo topic among women in the dressing room. Outside of the initial interview with management, asking another strip-club dancer, waitress, or bartender his or her age was taboo at any age.

If a young woman crosses state lines and chooses sex work, she is immediately a victim of a foreboding sort in the word of law—she is trafficked. But these cases seem excellent examples of how those legally categorized as trafficked can be agents of what others perceive as their "victimization." Such young women are dangerous employees for nightclub owners as minors, but especially as such minors who have crossed state lines. However, dancers are not technically employees of the club, and I was told in the 1990s that it was easy to obtain a plausible driver's license with a suitable birthday from other, older dancers.

The reality of how carefully such identification is scrutinized by management for what amounts to day-labor employment must be balanced with the supply-and-demand market for our bedazzled "widgets." Young girls are desirable as dancers. We heard this plainly stated on the Louisiana House of Representatives floor. In response to a recently passed New Orleans city ordinance with similar age restrictions, a state bill was proposed increasing the minimum age of strippers (but not other employees) across Louisiana from 18 to 21. On Wednesday, May 18, 2016, Republican second term member of the Louisiana House Kenny Havard filed an amendment which proposed to insert that strippers should additionally be

> between twenty-one and twenty-eight years of age, and shall be no more than one hundred sixty pounds in weight.

The bill passed without Havard's amendment, an amendment which can most kindly be framed as a bad sexist joke from a cis male perspective about over-regulation. Despite outrage by female members of the House, most vocally Republican Julie Stokes, the fact that such conversations occur within the discourse of the Louisiana legislature at all reflects a startling gender imbalance of power. The ease with which such sexist and ageist comments can be introduced demonstrates that the regulatory conversation about striptease dance in the French Quarter even today is discriminatory. It is unparalleled in any other industry which is part of the formal, tax-paying economy that official sessions concerning regulation should include such offensive comments about service workers. It is awkward for lawmakers to ask business owners to take precautions to protect vulnerable workers when they themselves behave like members of a bachelor party while engaged in making those laws.

Not only do the well-heeled tourists know that New Orleans is a welcoming destination; New Orleans is also a common destination for underage homeless people. Young people find many companions, ample abandoned or partially abandoned buildings where they can live as squatters, and a mild climate. Listening to women talk about their younger days on Bourbon Street there are many similar stories. Young women trafficked themselves; sometimes they contacted friends who were also on the run and perhaps mutually trafficked each other! In such cases I would agree with MacKinnon—that the law is constructed in such a way as to allow men to dominate women. In the case of the young B-girls I met, laws concerning adulthood, the age of consent, and migration across state lines were used by those who would continue to abuse (mostly) post-pubescent girls—whether by limiting their ability to support themselves or by assisting in their forceful return to hostile conditions in homes or foster care. Bourbon Street workers, in short, are in a much better position economically than young runaways elsewhere who engage in survival sex, exchanging sex directly for food and lodging. In my short, online and anonymous Wellness Survey, 4 of 11 respondents reported having experienced abuse at home before their work on Bourbon Street.

The chapters in this book reflect the different types of data that I use to document B-girldom. The data is placed in chronological order, with historic information preceding an ethnographic present of 2001. The last chapters are based on older women's experiences as they explained them to me during interviews in 2012–2016. This does not reflect the chronology of data collection—initially it was observations and interviews that led to historical probing.

I begin the book with a historic look at B-drinking in the city. I establish the long history of the practice, contextualizing my later ethnographic observations and interviews with a preliminary peek into the city's various archives. Written records expose that the system of B-drinking in New Orleans predates the existence of strip clubs in their modern form. B-drinking is recorded to have been happening in the city before the buildings housing the contemporary cluster of striptease clubs were entertainment sites at all; B-drinking is well documented to have been happening in clubs on nearby Canal Street when Bourbon Street was still a mainly residential area.

Chapters 2 and 3 reveal evidences in public records of the practice of B-drinking in the city's past and how it has changed over time. Archival data reflects the evolution of "B-girldom," a term I use to refer to the imagined person of the B-girl as she has been characterized in newspapers, novels, plays, and legal documents. Changes in the characterization of B-girldom reflect and are a part of economic and political shifts in New Orleans over the decades. This is a case study of the invention of the B-girl as a type, the evolution of an ideological character within the city into the public imaginary and hence into the "tourist imaginary" (Salazar and Graburn 2014) of the French Quarter.

Later, I document the daily realties of B-girldom as expressed by women in my sample of interview participants who earned significant portions of their incomes through the practice in the 1990s. In chapters 4, 5, and 6 I examine the ethnographic material I collected during the 1990s, with a focus on power and resistance in controlling incomes and the environment within the strip clubs. Using my most recent interviews, I examine various ways of remembering women's experiences as B-drinkers. Through interviews, photographs, and informal conversations I document strategies they used to earn incomes and manage stigma. Just as tourists are drawn to B-girls because of their role in the constructed tourist imaginary of New Orleans, so too do B-girls engage in image fashioning that reflected their own imaginaries of Bourbon Street. One important imaginary of the B-girls themselves reflected on a nostalgic, idealized past when Bourbon Street performers were celebrated entertainers known far beyond the local strip.

My data is most updated by two participants who were successful dancers in the 1990s and were working as "house moms" in 2014. House moms work as a sort of manager in the dressing rooms of large strip clubs. One was employed by a club on Bourbon Street, and one by a large, national chain of strip clubs and erotic material in another large city in the southeast. These women introduced me to younger women who are currently dancers, allowing me to continue the research. The breadth of data I have collected, I hope, allows for a more complete understanding of the sexual economy of tourism as it affects the lives of women engaged directly in tourism-oriented sex work in New Orleans and perhaps in other port cities which have converted from merchant hubs to tourism playgrounds.

NOTES

1. I follow the suggested terminology of the National Center for Transgender Equality (2014). Several of my B-drinking informants in the 1990s were transgender women, individuals whose gender identity was feminine but whose assigned sex at birth was male. None of these informants were cross-dressing men, as all lived full-time lives in which their identities and behaviors conformed to feminine gender roles. None had undergone sex reassignment surgery but did engage in hormone replacement therapy to develop feminine secondary sexual characteristics.

2. On post-Hurricane Katrina Bourbon Street, all three of the clubs where I had made observations and done interviews had closed or changed ownership.

3. Simply put, feminist jurisprudence is an approach to understanding the construction and the uses of law from a perspective that attends to its masculine assumptions, gendered implications, or gendered uses.

4. More recently, in explaining my research to a dancer in Wichita, her eyes lit up and she said, "Oh! You mean a Dancer's Screwdriver!" She was referring to the local practice of bartenders making "virgin" screwdrivers, or orange juice with soda water, for striptease dancers to trick clients into believing they were drinking alcohol. Although I have not researched sex work outside of New Orleans, such anecdotes cause me to suspect similar practices occur in many other parts of the United States. Bar girls are common phenomena worldwide.

5. http://www.creolequeen.com/creole-queen.html.

6. http://whc.unesco.org/en/faq/40

7. There are other areas of the city, such as hotels on Airline Highway, that are sites for sex work but well off the map of the tourist. These sites serve local customers and would not be considered part of the international sexscape. They are also places where the pricing for sexual labor is considerably lower than the French Quarter.

8. Judith Hanna (2012) does an admirable job of framing the discussion of striptease in a way that protects such performers. In her analysis of striptease and lap dance as a legitimate form of dance, she takes the anthropologist's perspective of relativism to include dancing relativism. Hanna's body of work builds on the notion that dance is symbolic human communication, which brings it firmly under the domain of the First Amendment and freedom of expression. Problematizing the discriminatory ways that nudity, self-touch, and performer/audience distance are regulated in strip clubs versus other local dance performances, Hanna makes clear that it is the message of the dance that is being silenced. She argues convincingly that legal limitations are placed on striptease that are not placed on other commonly consumed forms of American entertainment because of strong pressure from a powerful minority, the Christian Right. In legal cases, those who would limit the activities of strip clubs call upon what Hanna calls "the myth of adverse secondary effects." Hanna explains that court decisions based on the intent to protect the public have been formed on the misguided notion that strip

clubs attract "crime; spread disease; degrade women; deprecate property; and hurt a residential or business district or city image" (Hanna, 2012 [location 296/7421 of Kindle version]). Hanna builds on the literature that fails to support such a relationship (Paul et al. 2001, Linz et al. 2004, Hanna 2008a).

My work does not detract from such gains in the American courtroom, which protect dancers from being silenced. My focus is not on striptease dance. The practice of B-drinking I am dissecting here predates the contemporary clubs and is therefore obviously in no way "secondary" to them. It is also unfathomable that such an argument could be applied in New Orleans, where the city image sold by the local tourism economy (re)produces desires based on consumption, excess, and nudity.

Chapter One

Why B-drinking Works

A Semiotics of Seduction and Drinking

With skill and practice on the part of the performer, B-drinking can be completely invisible to a nonparticipant within the touristscape. The French Quarter is marketed as a place to drink, and memories of social drinking experiences are often shared with me by people to whom I mention my interest in New Orleans and tourism. To understand B-drinking, we can begin by critically examining the role of gender and social interactions in urban North American barrooms. Women in bars, whether as patrons or employees like B-drinkers, dancers, bartenders, or servers, are commonly offered gifts of drink. It is an accepted way to initiate sociality. To understand B-drinking and its risks, we need to think about the practice of the gifting of drinks by men to women in bars generally.

Joseph Gusfield (1996, 57) claims that shifts in drink, such as the time to drink coffee versus the time to drink alcohol, help to frame specific periods of work and leisure, respectively, across both industrialized and postindustrial cultures. For tourists, the shift in frame from everyday life to travel includes a shift in expectations about alcohol consumption as a part of a wider shift into a phenomenological experience of leisure. The experience of travel, both business and leisure, is generally marked by a change in consumption patterns broadly. Certain touristic practices, such as ordering room service, develop to meet these new consumption needs created by travel. Likewise, ordering a round of drinks for oneself and a B-drinker marks a shift in interactions between the B-girl and customer in the direction of greater intimacy.

New Orleans' reputation as a place to eat and drink freely contributes to its success as an enjoyable convention destination. Formal business can conclude, and workers can transition to less formal networking and socializing at a myriad of drinking and dining establishments of international renown. In a B-drinking interaction, drink is used to symbolically shift from a frame of stranger interaction to a frame of potential intimacy which would allow displays of sexuality, whether verbal, through body language, or through physical contact. Buying/receiving a drink from a person of the opposite gender is meaningful. As a tourist on Bourbon Street, buying a drink for a B-girl creates a tacit contract of exchange of the price of the drink for an experience of hospitality and conversation over a measured period of time with one of a wide range of enticingly dressed women.

The work of a B-drinker is first the creation of a sexualized identity, and second, the performance of emotional labor of an intimate and erotic nature with individual customers who purchase drinks in a bar. It is part of the occupation of the B-drinker to present herself as a sexualized, feminine object of the tourist gaze. John Urry (1990) defines this gaze as a way of seeing that focuses on meeting the expectation of the tourist, focusing on the metaphor of sightseeing. In New Orleans strip clubs, the tourist gaze is a sexualized gaze, and the frame of tourism allows those who would not go to strip clubs in their home communities to be comfortable doing so while engaging in the cultural practice of tourism. The B-girl, however, knows that her performance goes beyond appearance, and she uses perfumes, clothing textures, choices in music, and shifts in her manner of speaking to engage and entice customers. Even for women who do not allow any touch, the experience is more than visual. B-girls engage in emotional labor which includes not only performing certain emotions but controlling any contradictory emotions that might arise in an environment of ambivalence in which the sexualized body is expected yet verbal critique and sometimes abuse from customers is not uncommon.

Being a bartender in any setting, particularly in a tourist bar, is an excellent example of emotional labor. It is also a site where emotional labor can be challenging because of gender and different expectations of what kinds of interactions are acceptable performances to the tourist, who is him- or herself operating within a frame of leisure. Emotional labor requires the worker to control the display of her feelings in order to prompt in the point of view of the customer a frame of mind that encourages successful, ongoing transactions. B-drinkers allow the visitor to purchase a *performance* of interacting with a woman who sells sexual labor, whether verbal or physical, within a tourism frame of leisure outside of everyday life. Vessey and Dimanche (2003) refer to Bourbon Street as a "Disneyesque" enactment of nostalgia for vice, but surprisingly fail to find actual sex workers. If McKercher and Bauer (2003) are correct that tourists seek purchasable versions of the culture in places they visit, then the financial success of B-drinking as a strategy for young women in the 1990s suggests it fulfilled the tourists' fantasies about

New Orleans as an imagined place and provided enjoyable interactions at some level with sex work in New Orleans.

Good service requires the management and appropriate performance of the B-drinker's mood (or at least the appearance of their mood), which can lead to conflict in such nightclubs. My interactions suggest that the emotional management of exchanges between customers or employees ultimately helps contribute to strong feelings of community among employees who must work cooperatively. Especially when B-drinking is paired with topless dancing, women must provide "service with a smile" in the face of stark rejection and explicit critique of their bodies, personalities, or life choices. This vulnerability led to some very intense bonding between women in the Goffmanesque "back stage" of the dressing room.

Outside of the tourist encounter, the social script of men purchasing drinks for women is a part of heteronormative courtship in the United States very broadly, making the B-drink hustle a readable "offer" of emotional labor on the part of the tourist. A new customer quickly understands this way of structuring payment for individual attention from entertainers. In discussing her past work as a striptease dancer and B-drinker, Tish told me that

> A lot of places I've worked that didn't pay me to drink, I *still* got the customers to buy me drinks. It was the only way I could drink without paying an arm and a leg for it! And I think the customers don't realize there is any of this B-drinking going on because every strip club, all over the country, even with no B-drinking (commission) girls still get you to buy them a drink!—Tish, interview, 2014.

The difference between the B-girls hustle and a normative American script of seduction through intoxication is that the performance of a B-girl is commoditized labor; her control of her emotions and her ability to manipulate and control a seduction event translates to a wage. The customer knows he is paying the bar. What goes unsaid, explicitly, is that he is paying the *woman* for her time and attention.

At the heart of this ethnography is alcohol. Although I initially was drawn to the economy of Bourbon Street strip clubs expecting the most important commodity to be nudity, I found instead it was the company of a woman *while* drinking. All the clubs investigated were licensed liquor establishments, and none served food. B-girls and B-drinking laws are found within alcohol licensing requirements. Both customers' and B-girls' activities increased the alcohol sales that were the formal business of the three clubs where I made observations.

There are several very different types of drinking events that occur in the club and each is a dissimilar sort of social event from the perspective of the employee. In addition to scripted events of celebratory drinking such as bachelor parties, there is also a daily pattern of ritualized drinking among employees, particularly at the beginning and end of each shift. Both are structurally separate from any B-drinking event, which serves different purposes and follows a different pattern. The variety of patterns of drinking that occur in the French Quarter clubs *as a part of employment* exists alongside the more visible acts of drinking that occur as a part of the pattern of alcohol consumption normative to American leisure culture.

In 1987, Mary Douglas edited a volume addressing the question of alcohol from a "distinctly anthropological perspective." She points out that anthropologists have tended to ignore alcohol's social meanings, and medical and sociological studies that focus on the abuse of alcohol exaggerate its problems because of their Western bias. Included is an essay by Dwight Heath, proposing that it is a Western research bias that labels alcohol consumption and drunkenness as pathological, ignoring the many social contexts and functions of drink in cross-cultural context.

> The association of drinking with any kind of specifically associated problems—physical, economic, psychological, social relational or other—is rare among cultures throughout both history and the contemporary world. (Heath 1987:46)

According to the Social Issues Research Centre (SIRC 1998, 6) all societies consume some intoxicating substance, the most common being alcohol. It is not universal that the drinking of alcohol in social settings, or even the occurrence of events that encourage drunkenness, results in incidents of problem drinking. Indeed, there is ample evidence that responses to drinking and drunkenness are culturally mediated.

Although the consumption of alcoholic beverages is widespread, the relationship between specific societies and ethanol varies in meaning and construction. Fermented sugars are a naturally occurring substance that has appeared as an aside in ethnographies from around the world, but Douglas points out that alcohol itself does not become a topic of specific interest until the 1970s. Like tourism, alcohol comes as a recent addition to the anthropologists' interests. Further, studies of drink and drinking customs have been relatively ignored in favor of a social-problem orientation that frames alcohol consumption as pathology rather than custom (SIRC 1998, 4). Thus, it is not only in the ethnography of striptease that drinking has been an invisible but important component of social behavior. Rather, like striptease and indeed tourism itself, alcohol customs were ignored as frivolous topics not worthy of comment by the anthropological community until the second half of the twentieth century.

Douglas suggests drink is constructive in three ways and uses these functions to divide the contributed chapters to the edited volume into sections. "Drinks Construct the World as It Is" provides various examples of the role of drinking specific types of alcohol at specific times in different communities. Drink, like food more broadly, is not simply a material to be consumed. Beyond a materialist approach, food can express status (Fox 2014:3), identity (Mintz and DuBois, 2002; Ku 2013), and gender (Counihan 1999 and 2004). Michael Dietler and Brian Hayden (2001) examine feasting and its role in power ethnologically, and volumes are dedicated to broader semiotic approaches to understanding the diverse meanings behind food, translating the meanings of production and consumption in social and cultural contexts (Coleman, 2011; Brulotte and Di Giovine 2016).

Levi-Strauss (Elementary Structures of Kinship) asserts that "if the occasion required certain traditional foods, their very appearance, by a significant recurrence, calls for shared consumption" (1969:57), using as one of his examples "a bottle of vintage wine," and refers to the consumption of what should be socially shared as a kind of "social incest" (1969: 58). The type of drink (here vintage), the context of the drink (alone or in a group), and the timing of the drink can all serve as signifiers in a semiotics of drink. Different drinking events have and construct different meanings and are phenomenologically dissimilar experiences for the drinker.

Likewise, within the community of Bourbon Street strip clubs, there are distinct experiences of drinking. Drinking events between B-drinkers at the start of shifts help create solidarity, which is an important topic in the ethnographic section of this book. When I flip through photos of women from The Den, the club where I had the most rapport with the community and could produce and collect photos, one of the most common themes is that of groups of two or three women crowded around a shared drink smiling. In none of these photos is there a customer or a man in sight. These drinking events are not B-drinking, they are social acts of transition from leisure to work; in the case of the B-drinker, alcohol is an integral part of both leisure and work, but each drink has a context and a meaning.

Most drinking events within the club do not create a real solidarity, although ironically this is exactly what the customer seeks. The purchase of the B-drink initiates the beginning of work for the B-girl, solidifying the purchaser's role as customer, not friend. B-drinks tend to further distance the women from customers, especially customers who arrive within their own small communities. It was common for strip-club employees, from owners to wait staff, to complain about the attendees of conferences or other events when they arrived in large groups.

Another way Douglas acknowledges drink as constructive is through its importance to the informal economy in some illustrative communities. B-drinking is obviously an example of this type of Douglas' "constructive drinking." The sale of alcohol as a B-drink is the B-drinkers' work and is representative of the attention they are selling, time passage, and appropriate timing of payment for their services in some measurable, incremental unit.

Some women kept drink stirrers or swizzle sticks in their purse, a material reminder of the intangible commodities of time and responsiveness they had provided during their work shift.

Most useful to understanding the social phenomena of B-drinking is the first of Douglas' list of types of "constructions" through drinking. The social constructions and functions of the three different kinds of drinking found in my study site—employee/employee drinking, customer/employee drinking, and customer/customer social drinking—are invisible to the uninitiated onlooker but starkly clear to the worker on Bourbon Street.

According to the Social Issues Research Centre, drinking cross-culturally is always bound by culturally significant rules, "often the focus of strong emotions" (SIRC 1998, 7). The same report suggests some near universals about drink cross culturally, such as: there are always social controls on drinking behavior, social drinking is prescribed and solitary drinking proscribed, and there are *ubiquitous restrictions on female drinking* [emphasis mine]. In a review of possible mechanisms for such differences, Marja Holmila and Kirsimarja Raitasalo (2005) conclude that gendered differences in drinking have been found in all societies studied to date.

The Social Issues Research Centre suggests that this may reflect a biological constraint due to sexual dimorphism. Simply, there are physiological differences between males and females that can reasonably explain this agreement across cultures concerning gender and disproportionate prohibitions on women drinking alcohol. Thus the B-girl as a professional drinker is perhaps not only in violation of a local cultural norm, but indeed of a near universal culture of gendered human consumption patterns. B-girls violate many of the expectations of social drinking for women in American culture.

Gusfield emphasizes the meaning of drink in the routinized scheduling of time into productive time versus leisure time, whether by time of day (to show one is "off the clock") or day of week (first in Douglas 1987, 73; later reprinted in Gusfield 1996). Likewise, anthropologists of tourism have aimed to understand leisure time as framed in opposition to work (MacCannell 1976; Nash 1981, 1996). Gusfield categorizes the weekend as "a release from the rules and the tasks of daily routine" (73). Reminding us that in preindustrial communities work time was defined by natural limitations of daylight or individual corporeal energy, Gusfield states that social organization in industrial and postindustrial communities *create* the notion of "leisure as a definite and bounded part of time" (73). Industrial society introduced not only a change in the work people did but also in the way they organized their time.

Gusfield (75) hypothesizes that alcohol serves symbolic functions for Americans to pass from the two opposing behaviors of work and leisure. If drinking alcohol is a part of the experience of moving into *leisure*, the *professional drinker* becomes particularly problematic. Modernity in the West, Gusfield notes, insists that "the hedonistic, the playful, the irresponsible, the non-serious not be permitted to enter the domain of the 'serious' areas of making a living and earning a livelihood" (75). Expanding on the finding that there are different cultural patterns to drink, the same is true for drunkenness and problem drinking.

> Cultures (such as the US and UK) in which alcohol is only used to mark the transition from work to play—where drinking is associated with recreation and irresponsibility, and regarded as antithetical to working—tend to have higher levels of alcohol-related problems. (Social Issues Research Centre 1998, 9)

Here I will return to the ritual drinking that begins every shift of dancers. The "Social and Cultural Aspects of Drinking" report to the European Commission (1998) finds the following about such instances, when drinking is a part of transitional rituals. "Cultures in which drinking is an integral part of the normal working day, and alcohol may be used to mark the transition *to* work . . . Tend to have lower levels of alcohol-related problems" (9).

Although I initially hypothesized that follow-up interviews with women previously hired to B-drink would include widespread reports of alcoholism, this was not the case. An anonymous Wellness Survey distributed to my informants, which surveyed the overall health of women in 2015 who had worked in previous years as B-drinkers, found few reports of alcohol addiction. It may be that, like other "non-temperance" or "integrated drinking-cultures," the social groups formed by dancers simply carry those characteristics that are associated with fewer alcohol-related problems than more "ambivalent" drinking cultures (7).

Within a shift, the first drink is an event that occurs between small groups of the dancers, waitresses, and other women (including transgender women), as they first emerge from the dressing room changed from their street clothes into their stage costumes. This shift drink is particularly important, and the women are especially loud and bawdy while they share it at the start of shifts when large numbers of good customers are expected based on monitoring of the convention schedule. Symbolically, this shared drink between club employees starts the shift. As suggested by the Social Issues Research Centre (1998), this drinking event serves the cross-culturally recognized use of drink to "facilitate social bonding" (8) in a "socially integrative, egalitarian environment." I have never seen men included in this drinking event, even when they were employees of the club, starting the same shift, and well liked by the performers.

After emerging from the dressing room together, perhaps waiting on a friend struggling with eyeliner or some other costume emergency, groups of women would share a shift drink, always potent drinks made of hard liquor combinations sucked up through separate straws in a common glass. Sometimes the drink was a race, and sometimes it was a purposefully slow, drawn-out moment. This ritual drink often seemed like camp, an over performance of the gendered role of a feminine consumer. In her classic "Notes on Camp," Susan Sontag states:

> Camp is a certain mode of aestheticism. It is one way of seeing the world as an aesthetic phenomenon. That way, the way of Camp, is not in terms of beauty, but in terms of the degree of artifice, of stylization. (1966, 20)

Various performances of "play" (sexual, competitive, or sisterly) could be expressed by a group facing specific expectations about an evening's impending workload and remunerations. In a place where performances are for sale, this shared drink is a performance about that performance. The associated phenomenological experience of the rush of a "shot" style drink certainly adds to the potential for this drink to serve as part of a transformation from play to work, or perhaps from the real to the performed.

Girls in The Den, a decades-old topless club in a heavy foot-traffic area of Bourbon Street, preferred and later recollected with fondness sharing Mind Erasers, a drink made of vodka, a flavored liquor, and soda water. This drink is traditionally consumed as a single shot without soda, or through a straw with soda water, to induce a quick alcohol rush. When women drank shots as their shift drink, the bartender would shake the drinks in one large mixer, pouring them out into a row of shot glasses for the dancers to drink together. The Mind Eraser as a choice perhaps linguistically reflected a purposeful attempt to disassociate from any negative feelings about the coming work experience. It's interesting that bartenders interviewed ubiquitously agreed that the Mind Eraser was a "girl's drink." Of the many drinks shared as shift drinks, informants 20 years later universally recall the Mind Eraser.

The only thing that the shift drink has in common with other drinking events within the club is the presence of ethanol. This ritualized drinking event occurred two or three times between 7 and 8 pm in all three nightclubs where I made observations, as small groups of women leave the all-female, protected space of the dressing room for the public place of the club floor. Similar drinking practices can be observed among musicians, as well as among the waitstaff of many bars across the city. Some bars, in a nod to policy that does not allow workers to drink during their shift, allowed employees to arrive 10 minutes early to have a shift drink. Such drinking serves functions of social integration and perhaps the temporary leveling of all employees as equal players in the club for the duration of the shift. I suggest such drinks serve to draw new employees into the comradery of others working that shift. A "rush," or a feeling of quick intoxication, is part of the reason for this particular drink. This phenomenological reality, the physical experience of sharing the effect of the drink, serves as part of a rite of transition that assists the dancers, waitresses, and other women employed by topless nightclubs to shift experiential frames to that of work in an environment in which cooperation against perpetual sexual harassment is critical.

In contrast, B-drinking with customers has very different rules and expectations. One obvious difference is the desire to or not to feel the effects of the drink. In The Den, the drinks that were ordered during acts of B-drinking were limited by explicit club rules which required low-alcohol-content drinks such as champagne or even "champagne cocktails," a mix of champagne and soda water. Some bartenders are experts in creating a "float" of champagne behind the cover of the bar. A float is a thin film of bubbly wine on top of a glass of mostly soda water, which serves to reassure the nosy customer who insists on tasting a B-drink without requiring heavy drinking of the successful employee.

It is not incongruous that those who begin their shift with shared Mind Erasers often dump, dilute, or otherwise avoid consuming their "B-drinks." The logic is clear when one parses the different drinking events into separate semiotic and phenomenological experiences from the perspective of the drinkers. While the Mind Eraser is about companionship and a feeling of community, the champagne cocktail is consumed as part of a performance of mock-intimacy that is a commodity in the tourist industry. The Mind Eraser is an act to reify membership in a sort of voluntary association group, and the physical effect of the drink is part of the transition from leisure to work; consuming (or appearing to consume) multiple bottles of champagne while fending off any unwanted contact from amorous clients is solidly within the range of work.

CONSTRUCTING THE HETERONORMATIVE NORTH AMERICAN BARROOM DRINKING ENCOUNTER

The notion that drinking behaviors defined encounters, "framing" them in particular ways is evident in the literature on drink. Understanding the meanings behind drinking acts and responding in appropriate ways are an important part of Gusfield's notion of the "competent drinker." Gusfield (1987) suggests that comprehending the ways one is expected to behave within the "frame" (Goffman 1974) of a particular drinking situation makes one a "competent drinker." A competent drinker understands and follows the social rules for drinking in specific drinking situations. Gusfield suggests that those fluent enough in the semiotics of drink to be competent drinkers are more acceptable than those who choose not to drink at all, which is viewed as antisocial. There are expectations of competent, gendered public drinking that B-girls understand and purposefully manipulate. This leaves them at risk as targets for men's aggression, a noted component of "drunkenness" that, as we shall see, is scripted for North American men.

Jacqueline Ferris (1997) finds that men and women experience drinking and dating from very different perspectives. In a qualitative analysis of interviews, she discovered a gendered patterning of themes among participants describing public drinking events. In summary, Ferris (1997) found that while women focused on situational control through *risk aversion*, men focused on situational control through *manipulation and aggression*. While Ferris aimed to understand the role of alcohol consumption and aggression in dating situations, I would argue that the B-drinking encounter is successful precisely because it is modeled on this normative script of gendered public drinking. The B-drinker is employed to engage in the emotional labor of behaving as a competent drinker in a setting where her patron (the customer) hopes she might become drunk, hence becoming vulnerable to his manipulation and potential aggression.

Ferris (1997, 668) muddies Gusfield's (1987) idea that drink serves generally as a signifier for framing situations by introducing such deeply gendered patterns in drinking behaviors. Drinking can indicate a shift in frame from business to a more casual meeting; for example, the introduction of wine (or food) at a late meeting for white-collar workers can indicate a shift in frame. A semiotic domain parallel to the sartorial "loosening of the tie," the introduction of alcoholic drink to a Friday meeting or a long lunch serves to shift out of a purely "work" frame. This may allow for the lengthening of a work day through an illusion of casual interaction, such as occurs during the business lunch. This shift in structural time through drinking acts is not the same for men as women, as they are expected to manage social control while drinking in disparate ways.

In "Courtship, Drinking and Control," author Craig Reinarman (1988) interviews young Canadians about the symbolic meaning of drinking and drunkenness in the context of dating. Reinarman proposes that women are more likely than men to believe alcohol impairs one's ability to control one's behavior. He suggests that the sexual double standard leaves women more open to risk if their drunken behavior included stigmatized behavior. Thus, women are more invested in being able to separate their drunken behavior from their "true selves," and benefit more than men from a *belief* that they lose agency with drink. Although drunkenness and its believed loss of inhibition is antithetical to North American femininity, which requires control, it also can be mobilized by women as an excuse for their behavior.

Broadly, Reinarman (1988) focuses on the dual role of alcohol. Ethanol consumption by humans has physiological effects but also creates a socially sanctioned frame of drunkenness which the drinker can use to act in culturally defined ways (such as promiscuous or aggressive). This cultural definition of drunken behavior as a part of sexual interactions is certainly modeled across multiple media representations of romantic encounters for Americans.

Among Ferris' (1997) research participants, limited to heterosexual adults without children, she found qualitative differences in the interpretations of drinking behaviors between men and women when engaged in social events. Particular drinks and drinking styles were reported to be influential in people's decisions about the drinker's desirability as a mate. Identifying categories of themes within the interviews, Ferris found that for women their key category was risk limitation, while additional important categories included altered reality drinking, symbolic communication, learning to drink, and sociability. Men's responses had little overlap with women's. Although symbolic communication was also a common theme, escape drinking and control were the other major categories of discussion for men. In the end, Ferris finds that the drinking/dating script was focused on control for men and on risk limitation for women. In this sense, B-drinking is broadly similar to other forms of barroom drinking in the West; B-drinkers are professionally sociable while drinking and must constantly be aware of potential risks from clients who hope for control over the situation.

Ferris' female informants strategized during drinking events to be sure that they would be able to get home safely and suggested that accepting a ride home from an acquaintance "sets up expectations, framing the situation as a sexual one" (1997, 680). Such expectations, women reported, left them at risk of violence. Meeting new men in public spaces such as bars was perceived by women as unsafe, and female informants felt that drinking made them more vulnerable to unwanted sex. According to Ferris (1997, 683) "Several women . . . said they had had unwanted sex to avoid hostility or the likelihood of a violent attack." As one of her informants explained, "They kind of expect that you're going to do something with them, and you've had a few drinks, and it's kind of just easier to [do it] rather than try and fight them off and say no . . . " (Ferris 1997, 682, informant 115/9). Again, this is consistent with the experience of B-drinkers in the 1990s who were sometimes aggressively pursued by customers with whom they had consumed a series of drinks.

While Ferris found that women perceived social drinking as risky and potentially leading to unwanted sex or violent attack, men perceived drinking encounters as potential sites for "initiative taking, situational orchestration, and aggression" (1997, 683). Rather than risk control, men sought full control "over themselves, over the situation, and over the women they were dating" (683). Ferris reports that men recalled times when they felt manipulated or powerless in a dating situation as examples of "worst-date scenarios." Men ubiquitously agreed that they did not prefer women who initiated contact, although they tagged such statements with qualifiers about their personal "traditional" values. Men also reported that drinking alcohol increased the quality of their ability to present a positive version of themselves. "Basically, I'm more myself when I've had a drink . . . " (1997, 686, informant 208/20). Drinking just enough but not too much was a way to display control, the ability to perform as a "competent drinker."

Overall, Ferris demonstrates that for her informants, drinking holds specifically gendered meanings in the script of heteronormative courtship. For men, manipulative tactics are meant to initiate sexual contacts that women as gatekeepers are expected to resist. A woman who resists too little is undesirable. A woman who resists too much faces violence. A woman who resists but allows a man to control the situation through persuasion, seemed to have been the ideal. Ferris makes clear that this familiar script, absorbed by both men and women, leads to interactions where men who lose control and become violent because of women's risk aversion strategies are viewed as responding *normally* to drinking and drunkenness in the context of heteronormative courtship. To avoid aggressive responses, women have to be careful to not be identified as a "cock-tease" (Ferris 1997, 694, informant 202/13).

Returning to the B-drinker, it is in fact her *explicit goal* to be a "cock-tease," exactly the behavior Ferris' informants reported as creating high risk to themselves. Returning to a tourism-studies perspective, Jon Donlon and Jerome Agrusa insist that similar to other such "club clusters" (2003, 121), Bourbon Street is reported to be "powerfully regulated and virtually entirely free of its authentic rough, sailor's bar, sleazy joint, harbor-front historic reality" (121). Thus, nothing *can* happen except the tease, from a formal, top-down perspective of the industry. It is the tease that protected the interests of the "venture capital injections" from 1975 until 2000 which "updated and upscaled" Bourbon Street "to reposition them in the competitive hospitality marketplace."

The B-drinker is part of a performance of heteronormative drinking rituals, but inverts (or negates) the expectations of control over the situation by failing to lose control over her sexuality as she continues to accept alcoholic drinks. Certainly I have observed many cases when the expectation for sexual intimacy based on a woman's attention over drinks in a bar lead to frustration for the customer purchasing B-drinks (this is why such nightclubs always employ bouncers or male management). The expected lowering of sexual inhibitions with drink on the part of the woman does not occur in B-girl interactions with customers, even after the apparent consumption of as many as 3 bottles of champagne. If Ferris is correct, and alcohol provides North American women with a frame within which otherwise desired sex can be more legitimately pursued, it is clear that the subculture of B-girldom provides a very different script. What is legitimized through drunkenness within the dressing-room community is very different from the apparent American feminine norm, based on the above reports.

For the B-girl, rather than drink leading to vulnerability, it leads to empowerment through economic gain. Vulnerability is also moderated by the group environment—the customer is not in a position to control the interaction in a strip club, particularly if he is a short-term visitor to the city who is unlikely to be a repeat customer. It is perhaps not surprising that such an inversion leads to possibly the most aggressive of acts—the creation of a law that restricts such behavior among women. B-girls, specifically (as opposed to potential B-boys) defy the scripted norms of gender, courtship, and alcohol. As we shall see, the Louisiana State Legislature persisted for many decades in maintaining that a gendered law concerning "B-girls" was a response to a gendered problem and hence not discriminatory in nature. Laws against B-drinking have also not been written and maintained to protect women against perceived vulnerabilities to addictions or other harms in the workplace. Charges against B-drinkers cited in newspapers and courtrooms instead commonly center on complaints about the low alcohol content of the drinks, or the intention of the woman in accepting the gift of a B-drink to trick vulnerable men. Indeed, in a lengthy search of archival documents I have found only one suggestion of a case in which a lawmaker suggested that attention must be paid to either the ethics or health implications of paying women to consume intoxicants for their livelihoods. B-drinking laws were written to protect men from women.

The relationship of alcohol, sexuality, and control is of central concern in the last chapter of this book. Based on my sample, B-drinkers are much more vulnerable to homicide or attempted homicide than women who are members of the general population. Many of these events seem to occur between the B-drinker and men with whom they have personal relationships. Ferris suggests that her study offers insight into aggressive behavior as a response to frustration based on role expectations (1997, 700). Thus, in addition to bearing the stigma of being framed as criminals, women who violate the scripts of drinking behavior in bars in heteronormative culture sometimes risk torture and death. Men behaving aggressively toward women were common events in the clubs, particularly women perceived as being surprisingly in control of their decisions despite their appearance of drunkenness. Women who were more masculine than feminine in their drinking patterns were more likely to suffer from male aggression, sometimes through manipulation and sometimes through brute force.

Chapter Two

B-girls in Public Discourses

A Heritage Approach to Bourbon Street

The practice of soliciting patrons to purchase drinks for pay is part of tourists' experience of Bourbon Street today because the strategy has long been an integral and accepted part of French Quarter life. The B-girl as a journalistic and literary figure has been a highly visible component of the imagined French Quarter since the mid-twentieth century. The B-drinker and her scripted hustle appear in many newspaper stories and famous narratives of New Orleans as a normalized phenomenon; like other French Quarter hustles it is generally tolerated as an expected, humorous, and logical part of business on Bourbon Street.

This is particularly so in portions of the French Quarter which are a part of the contemporary American sexscape, such as the strip clubs of Bourbon Street. Denise Brennan's notion of the sexscape (2004, 16) as an imagined map of sites which draws international travelers for paid sex, and which is marked by inequalities of power between those performing erotic labor and those purchasing it, is applicable in understanding the allure of New Orleans. Bourbon Street is one node on the contemporary American sexscape, virtually synonymous with strip clubs and drunkenness for leisure and convention tourist alike.

> The classic instance of the Quarterite's placid acceptance of the unacceptable was the case of the berserk client in a Vieux Carré nightclub who almost wrecked the place in a fit of pique. And the patrons thought he was part of the show.
>
> It seems that the client had been sluffing off his spondulics right royally in purchasing B-drinks for one of the hostesses, a B-drink being water impregnat-

> ed with vanilla essence which the swains obtain for the girls at perhaps $1 a throw in the hope of winning their affections.
>
> The B-drinker, overcome by the generosity of the swain, agreed to tiptoe through the tulips with him come closing time. After his bankroll had been exhausted, she suddenly excused herself, stating that she had developed a headache and was going to take a powder . . .
>
> An hour later, when it dawned on the swain that she had played him false, his wrath was fearful to behold. (Jacobs, *Times-Picayune*, New Orleans, Dec. 26, 1948)

The prototype of a clever, hustling B-girl is a part of New Orleans historic past, here represented in *The Times-Picayune* in 1948; she is a playful criminal, viewed as a typical way that tourists, soldiers, and other travelers might engage with vice. By the 1990s, the successful B-girl performed emotional, and sometimes sexual, labor in such a way that the guest felt temporarily pulled into the back scenes of tourism as they imagine them to exist.

A historical look at the construction over time of the New Orleans B-girl in the mediascape is at the heart of this chapter. In Arjun Appadurai's mediascape, globalization as felt through the social practice of imagining is found in the newspapers, cinema, plays, and political narratives whether in news or legislature; in each of these sites the B-girl exists and is created. Such constructs of Bourbon Street have long been consumed by New Orleanians and visitors, revealing the ongoing social imagination of the B-girl. In fact, she is in part created through such narrations.

Investigating historical narratives of B-girls as a character type helps to reveal the invention of "B-girldom" as phenomena in New Orleans and in Louisiana more broadly. I borrow from philosopher Ian Hacking (1986) the notion of the cultural creation of categories of workers, elucidated in his piece "Making Up People." Building upon Michel Foucault's notions of the history of knowledge, Hacking points out that it is only with the introduction of a *category* such as the "B-girl" (in this chapter in media widely circulated in New Orleans, and in my next chapter in the legislature) that it becomes possible for one to engage with B-girldom as a strategy in the reality of tourism work on Bourbon Street. He states, "Making up people changes the space of possibilities for personhood" (Hacking 1986, 156). Similar to his example of the French *garcon de café*, there are many touristed sites where people engage in the service industry in culturally defined ways. It seems fair to say that there are people who engage in service in every society, yet the ways they do so and the needs they meet create categorical types of person which it then becomes possible to be.[1] It is possible to work as a B-drinker only because of a historical and cultural niche, a sort of made-up person in the laws and narratives of the city. B-girls are an expected part of the landscape, real and fictive.

Peculiar cultural institutions create unique categorical types of person, and those distinctions powerfully organize reality. Like Hacking's (1986) example of the French *garcon de cafe*, B-girls become distinct through their classification as distinct; the act of defining B-girls in media and legislature is a powerfully creative act. With these ongoing narratives that craft the classification of B-girldom, the B-girl comes into being and changes with the times. B-girls, a sort of made-up person in the laws and narratives of the city, not only exist but have become an *expected* part of the touristscape. In both real and fictive stories of the New Orleans touristscape, the B-girl has been a named and well-known personage dating back to the 1940s. These narratives suggest that from the middle of the twentieth century until the present, B-drinking has played a role in the sexual economy of tourism on and around Bourbon Street. B-girldom provides a case study in how the intersection of tourism and sex work is deeply embedded in regional history.

NEW ORLEANS, TRAVEL, AND A CHRONOLOGY OF REPRESENTATIONS

The social construction of B-girls in popular media—novels, movies, and newspaper stories about New Orleans—reveals they inhabit an ambivalent space. Playful images of buying B-drinks appear in reports of an authentic New Orleans experience. At the same time, B-girls as subjects are sometimes portrayed as dangerous criminals. B-girls are both embedded in and a blight upon the face of the French Quarter, both authentic and illegal inhabitants of Bourbon Street. Constructions of B-drinkers in the public imagination play a critical role in controlling sex workers who are otherwise in positions of power in their service (or refusal to provide service) as dancers and B-girls on Bourbon Street as a heritage tourism site. Public narratives pair with laws against B-drinking to counteract the potential for collective action of such women, who as a group we will see can act as powerful stakeholders in the French Quarter.

The congregation of sex workers and visitors in the city of New Orleans is not new. Alecia Long (2005) details the historically deep links between travel to New Orleans and the city's sexual economy. The short-lived, legal red-light district of Storyville (1898–1917) was convenient for travelers, located at the end of the early passenger train lines just outside the French Quarter. Although reformers advocated for the closure of the Storyville district, Mayor Martin Behrman realized that the district contributed to the appeal of the city for tourists (Long 2005, 226–227).

Jazz historian Al Rose (1974) documents that B-drinking was already practiced in the saloons in and around the brothels of Storyville one hundred years before my interviews in the 1990s. The historic district was unofficially named after Alderman Sidney Story,[2] who wrote the ordinance forbidding prostitution anywhere in the city *outside* of the district. This created within the boundaries of the Storyville neighborhood (across Basin Street from the French Quarter, in part of the current Treme neighborhood) a sort of de facto red-light district. Story hoped that containing the brothels to one neighborhood would protect property values in other parts of the city and keep vice away from respectable women and families.

For a short time, Storyville was the home to a wide range of sex workers. Poor women in small "cribs," a simple rented room usually facing the street, sold sexual services cheaply. More upscale addresses such as those on Basin Street provided ritzy and expensive entertainment including live music, liquor, and expensive sex. Within the boundaries of the district, elaborate mansions were built by madams who became rich and notorious for running brothels that served the "sporting man" who visited New Orleans to attend the horse races.

Storyville is well documented through the photographs of Ernest J. Bellocq, whose striking portraits of Storyville women in casual poses became famous posthumously. Bellocq's work continued to influence the imagined New Orleans through the mediascape when director Louis Malle created the historical fiction film *Pretty Baby* in 1978, which tells the story of a young girl growing up in Storyville whose mother was one of his models. Some of Bellocq's photos have had the faces scratched away, and whether this was done by Bellocq himself or by his brother, a Jesuit priest, is a New Orleans mystery in its own right. The sixteen square blocks of the Storyville neighborhood were later demolished to build the Iberville[3] housing projects. Bellocq's photography, the famed "Blue Books" which advertised the brothels, and other archival remains serve as rich sources about the district. In the 1990s, daytime walking-tour guides of the French Quarter regularly stopped their tours on Basin Street to retell the history of America's historic legal district for prostitution. Storyville and legalized prostitution continue to be a part of the city's remembered heritage, actively marketed to eager tourists who listen while sipping mint julips at noon one hundred years later.

As early as the Storyville years, the strategy of B-drinking served as one way to initiate contact between women selling sex and their clients in New Orleans. In his book *Storyville, New Orleans*, Rose (1974) quotes from an interview in the 1960s with "Lew," a retired railroad worker. Lew had experienced being on the train as it passed directly in front of Storyville, just before the New Orleans station. He describes the scene when the train would pull in to the old station at the beginning of the twentieth century—"We'd [the train] come in real slow—and there'd be the girls standing at the windows . . . just posing."

> In the saloons they had these girls, too, mostly a lot tougher. They'd be what you call today, "B-girls." If you weren't *marked* (with a flower or a feather) one of these girls would come up to the bar and ask you if you'd like to go sit in a booth with her and buy her a drink. If you did, she'd have her drink with you, probably hers would have hardly any alcohol in it, and she'd try to get you to take her either to a house or sometimes upstairs in the cabaret. (Rose 1974, 162)

Here, we can infer cooperation among women involved in the sex trade, who used inexpensive sartorial markers, such as a lapel flower, to symbolically reveal to other sex workers those who had already purchased sex from those who were still potential customers. In Storyville as today, women followed a script of soliciting drinks while purposefully concealing their low-alcohol content as part of a frame in which to negotiate the sale of various types of sexual acts. It is difficult to know to what degree B-drinking supported Storyville through liquor sales, as the practice seems to have been yet unnamed.

Richard Campanella (2008, 294–296) cited the claim that in 1788 wine and liquor made up one-third of the imports into Louisiana, noting both the geographic positioning as a port and the southern European and Mediterranean Catholic influences of the city's population as variables important to understanding local attitudes toward alcohol.

> Sailors, travelers, visiting businessmen . . . liberated by their anonymity and decoupled from the responsibilities and restraints of home, gravitate to opportunities for immediate gratification, in which alcohol usually plays a primary role. (Sex is a close second—and, not coincidentally, accounts for another historical reputation associated with this city.) (Campanella 2008, 296)

In the character of the B-girl, we find the combination of these two most notorious of New Orleanian vices: alcohol and sex.

QUAINT CREATURES: BEGINNING THE DISCOURSE

New Orleans' most widely circulated newspaper, *The Times-Picayune*, is a place where narratives of B-girls abound, playing their own roles in the changing social milieu of the city. Changing images of B-girls between their first mention in 1941 and modern day reflect the broader political and economic shifts that have shaped New Orleans' French Quarter. As we follow the B-girl through time, her job stays the same, but her role changes. A review of the discourse illuminated by the New Orleans *Times-Picayune* reveals shifts in attitudes towards B-girls and B-drinking from 1941 to 2012.

In the post-Depression years, burlesque in New Orleans included a mixture of striptease, comedy routines, and variety acts (Milner 2012). In *Strip-*

tease: The Untold History of the Girlie Show (2004) Rachel Shteir suggests that the evolution of burlesque into striptease in New Orleans during the twentieth century does not follow the same trajectory as in other parts of the United States because of New Orleans' unique history as a "pleasure town." This is reflected in the role B-girls play in the scene on Bourbon Street. B-drinkers, as tourism service workers, fit into New Orleans' erotic industry particularly because they provide a liaison between visitors seeking pleasure and Bourbon Street as a cluster of potential experiences.

Here, I take a detailed look at local newspapers and locally produced or interpreted fiction to reconstruct the place of B-girls as a French Quarter icon and discursive category since at least the 1940s. The B-girl as a character was investigated through an exhaustive search of 14 terms (such as B-drinker, bee drinker, and soliciting drinks) related to the practice in *The Times-Picayune* newspaper. Between the period of 1941 and 2012, 293 articles were deemed relevant. Legal documents, such as from the state legislature, city ordinances, and appellate documents, served as additional resources and are discussed separately. Figure 2.1 indicates the chronological distribution of these relevant articles.

The B-girl first appears in the *Times-Picayune* as a "hotspot hostess" (Melancon, Feb. 2, 1941:2) focusing on her role as a trickster, plain and simple. This "Bee girl" was reported to change her name as she moved from town to town, and she is headlined as an outsider—"Juke Boxes, Dance Girls

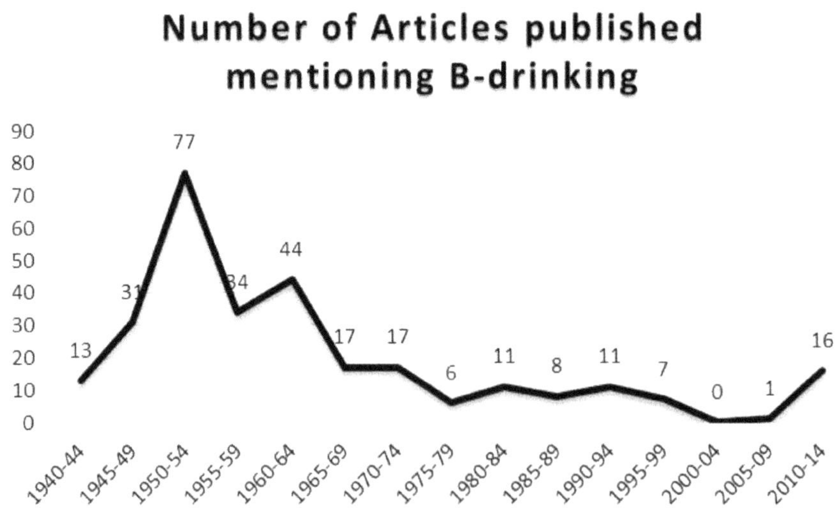

Figure 2.1. Distribution of 293 Articles Published Mentioning B-drinking Located in the Archives of *The Times-Picayune* Newspaper. Created by author.

Invade Bayou Country, Bee-Drinker Relates Tales of Her Hitch-Hiking Odyssey." Traveling with a male companion who helps her find her employment, she represents herself as dependent. In her self-derogating report, "I'm not smart that way. I get scared when I see a typewriter and seven plus eight is 12 every time I try to add it up" (1941:2). She was a woman whose "practice it was to induce customers to buy them imitation drinks which were merely colored water" (*Times-Picayune*, Feb. 11, 1943:23). Her earliest crime, it seems, was simply to sell nonexistent liquors.

In the early 1940s, the B-girl can also be found in the entertainment section of the Sunday paper. According to *The Times-Picayune*, the typecasting of Jean Phillips as a *dame noire* helped her in her quick success, described as "doing a sudden Cinderella." The suggestion is that Phillips looked like Ginger Rogers and was cast in roles that contradicted Rogers' public persona. "Jean's a meanie—hardly the Ginger Rogers type," said the paper, describing Phillips' part as "a small bit as a B-girl . . . (that) caused a sensation" in *Among the Living* (Nov 16, 1941). The full synopsis of the film by Turner Classic Movies reads that "Peggy Nolan, a slinky blonde, entices him into buying her a drink." It doesn't imply that Nolan is a paid B-drinker; this is read in by the New Orleanian local who interprets the film himself.

In a political cartoon published in 1943 (see figure 2.2), we find a complaint against increasing regulation at that time and its relation to World War II that uses the character of the B-girl. In the image, a woman in a long gown labeled "Bureaucracy" pours a customer a drink from a bottle labeled "Befuddling Multiplicity of Rules and Regulation," urging him to "drink up." In the background, a police officer questions the bartender, asking, "I guess you're too busy to know what's going on here?" Such imagery would not be understandable or funny to the readers of the paper if the B-girl did not already exist in the local imaginary.

All kinds of visitors have encountered B-girls in New Orleans, and B-drinking in *The Times-Picayune* parallels historical flows of visitors through the city. During the Second World War, Americans moved about a great deal, and as a port city, New Orleans always had a strong military presence. The intersection of sex work and the military industrial complex is well documented, as camp followers, military slaves, and base prostitutes followed the flows of money and men in the economy of war. Cynthia Enloe argues that "militarism and tourism . . . may be kin, bound together as cause and effect" (2000, 68). The laws concerning B-drinking, like the legal closing of Storyville, were related to confrontations with and disruption of the Louisiana military presence.

The increase in the number of soldiers visiting the port of New Orleans for rest and relaxation was significant in the early 1940s. The Louisiana Maneuvers, a series of U.S. Army exercises between 1940 and 1944, brought thousands of soldiers to central Louisiana who would certainly have spent

A B Girl Who Causes Lots of Headaches!

Figure 2.2. 1943 Political Cartoon Depicting a B-girl Serving "Rules and Regulations" Temple Cartoon Reprint—"A B Girl Who Causes Lots of Headaches!"—February 10, 1943.

time in the French Quarter and specifically on Bourbon Street. Perhaps equally important were the growing numbers among the working class of New Orleans employed by Higgins Industries. The building of LCVPs [landing craft, vehicle, personnel]—nicknamed "Higgins boat"—at its peak employed 20,000 people in New Orleans (Brinkley, 2000). The inspiration for the founding of the National World War II Museum (*Times-Picayune* Nov. 17, 2011), Andrew Jackson Higgins expanded his A. J. Higgins Lumber and Export Company into eight plants in New Orleans, which are credited with having built 92 percent of the American Navy vessels by September of 1943.

The first thematic set of articles, all portraying B-girls as unpatriotic, begin to appear in 1943. Although in September of 1941 B-girls were just "meanies," as the war continued they became actively harmful to the war effort. The discharging of "hostesses" was announced (*Times-Picayune*, January 28, 1943), and shifts toward bars that employed only non-B-drinking waitresses was touted as improvement. *Times-Picayune* authors discuss B-drinking in two distinct sections of the paper—news and the crime section. In 1944 and 1945, complaints were reported by armed services personnel (army and navy) against establishments employing B-girls. This brings B-girls into symbolic opposition with the military social order, a unique aspect of B-drinking discourse for that historical moment. This theme continues until roughly the end of World War II, with the last few sporadic references in 1945 as late as 3 months before the end of the war.

In 1943, B-girls were denounced as "a demoralizing influence," who "[cause] the soldiers in the city to lose their money" (*Times-Picayune*, Jan. 23, 1943, 1). Even the owners of clubs who previously admitted to employing B-girls deprecated the practice as anti-American. "Bee-girls for patriotic reasons are out for the duration . . . there won't be any drink soliciting in this establishment" (*Times-Picayune*, Feb. 2, 1943,1). Later that month, the chairman of the Social Betterment Committee, Holmes, was quoted saying that bars employing B-girls would "find it difficult to set themselves right with the patriotic and respectable public" (*Times-Picayune*, February 11, 1943, 1). Unconvinced that the practice was changing, Holmes responded that "changing clothes doesn't change the business."

There was some ambivalence within the population of the city as to whether it was more important to protect visitors or to protect the jobs of women employed in the bars that served them. By April 14, *Times-Picayune* writers mocked the protection of the public through this crackdown on B-girls, noting that other forms of vice remained legal. "The ancient art of bee-drinking has been abolished so that the military and citizenry can save their money. But just to make sure they don't save too much, the one-armed bandits have been reinstated" (*Times-Picayune*, April 14, 1943). Here, the author jokes about the crack down on B-drinking in contrast to gambling. Gambling, here specifically slot machines, is the vice that gives New Orleans dialect the historic term "sporting man," which was also applied to the customers of burlesque. On June 21 and October 4 of the same year, crime reports identify cases of women charged and arrested for "Bee-drinking," in one case including the home addresses of the women in question, obviously with consequences of stigma and vulnerability. City leaders working to control "vice" during this era considered "tightening up" laws controlling B-girls alongside those concerning prostitution, equating the two practices (*Times-Picayune*, March 29, 1944).

Like strippers in the 1990s, B-girls in the mid-1940s "sit in booths, and order double champagne cocktails, and . . . entertain male customers" (*Times-Picayune*, July 16, 1946, 8). The B-girl in the 1940s is portrayed as being the agent rather than the victim of her engagement and *Picayune* writers view managers as "*allowing* [emphasis mine] women on the premises for the purpose of soliciting alcoholic drinks" (*Times-Picayune*, Feb. 25, 1947,

21). B-drinkers of the 1940s earned high incomes for what may be considered semi-skilled labor. According to the case of "Ruth Lewis" (*Times-Picayune*, Oct. 24, 1946, 4), reported in 1946, B-drinkers were earning 50 cents per drink, or $8 for every quart ordered by a client.[4]

Defining B-girldom was a project of the 1940s, when B-drinking became prohibited by still (mostly) unchanged acts of the Regular Session of the Louisiana Legislature in 1948 under the Earl K. Long governorship. City ordinances regulating the licensing of liquor establishments (City Ordinance No. 15697) already banned B-drinking (*Times-Picayune*, Feb. 7, 1945, 1), but apparently were not addressing the problem directly enough. Specifically, the law states that establishments may not:

> Employ or permit females, commonly known as B girls, to solicit patrons for drinks and to accept drinks from patrons and receive therefore any commission or to receive remuneration therefor in any other way. (Regular Session of the Legislature of Louisiana, Act 360, Section 57 F)

With the penning of the law, the B-girl as a character became a legal category of persons. Like many social and political issues revolving around sex work,[5] the discourse begins in the law. In this case it is embedded in laws meant to control another vice—alcohol.

Often, newspaper men stopped to define the B-girl. Some simply quote the law; some discuss the ambiguity of the term in the law itself. The commonness of the term "B-girl" in the New Orleans dialect, as reflected by its use in *The Times-Picayune*, reveals the commonsense nature of the B-girls' existence in local New Orleanians' ideas about the community within the French Quarter and similar entertainment districts across the Gulf Coast. Journalistic definitions provide ethnohistoric moments that widen our understanding of the category as constructed and understood within the city. This clarifies B-girldom as a category, giving us a better idea about change and consistency in gendered performances of sexual and emotional labor in the nightclubs that line Bourbon Street.[6] B-girls, and similar characters of playful vice, shift at specific historical moments in ways paralleling contemporary themes in the discourse of sex work more broadly.

An early article speculating on the origin of the term begins with a benign explanation—perhaps the term reflects that many operate on Bourbon Street? Perhaps it is short for "bar girl?" Other suggestions are that they may have begun by ordering brandy drinks—the most expensive drinks in the house (*Times-Picayune*, Feb. 27, 1949, 2). In fact, in the 1990s B-girls often ordered expensive champagne drinks (or champagne by the bottle). More complementary explanations include the suggestion that they are "as busy as bees, and you might get stung" (*Times-Picayune*, Feb. 27, 1949, 2).

By 1948, a B-girl was understood as "one of those quaint creatures who entice men to drink so they can make a percentage on the liquor sales" (Hopper, March 26, 1948, 24). Quoting Merriam-Webster, *quaint* is defined as "having an old-fashioned or unusual quality or appearance that is usually attractive or appealing." The B-girl, by 1948, was already considered an old-fashioned institution of the French Quarter's nightclubs, hinting at the darkly obsolete, alluring, but somehow dated tradition.

NEW ORLEANS: THE UNITED STATES' "BANANA REPUBLIC"

After the war, B-girls were again free to fall into their less national, more local role. A seasonal piece, "Carnival King Reaches Mouth of River" includes a narrative cautioning the visitor about B-girls. Like the pilot of the boat bringing in King Rex and starting Mardi Gras, a B-girl is described as "a bar pilot. She pilots you to the bar, across the bar and to a table if you are not careful" (*Times-Picayune*, Feb. 27, 1949, 2). Around this time the discussion of vice and B-girls begins to be conflated with prostitution. Although B-drinking laws are and have been embedded in alcohol licensing laws, by 1946, the paper discusses "prostitution, gambling, and 'B' girl cases" in one breath—a dramatic shift from the previous decades where B-girls boosted alcohol sales and gave you nothing for your money.

Around 1948, the public discourse surrounding B-girls shifts in a malevolent direction. B-girls begin to be associated with crime and violent criminals. The B-girl moves from a nuisance to sailors to a murderess. Interestingly, the shift seems to begin in fiction before it appears in nonfiction news reporting. In March of 1948, "The Case of Millie Pearson," an entertainment piece about a B-girl who murders was written by Norman Reilly Raine for the *Saturday Evening Post* (Hopper, March 26, 1948, 24). By June 1948, a nonfiction piece ran about a B-girl who "lured" a man to her husband's home, where he beat and robbed the victim. The article indicates that the courts only charge against the woman was a charge that was held against all three—drunkenness. Her husband was charged with "disturbing the peace by assault" (*Times-Picayune*, June 24, 1948, 40). Again, on February 17, 1949, a woman identified as a B-girl appears in the paper in an article about her arrest for "running around with a fugitive" (Clinton, 1).

Discussion of B-girls takes a twist in the early 1950s, as language used to discuss the activity implies such women are not actual employees of nightclubs where they are arrested. Rather, articles suggest that they are "floaters," only in cahoots with managers or bartenders (*Times-Picayune*, April 13, 1950, 2). This provided a necessary separation between the city's elite and service workers; one very well documented case of using "knock-out drugs" to rob a client occurred in a nightclub owned by a member of the "vice committee to the mayor" (*Times-Picayune*, April 6, 1950, 12). Ethnographic observations of B-drinking in the 1990s revealed that part of the jobs of managers and bartenders (and a reason they received relatively high pay in comparison to other bartending jobs in the city) was to go to jail in the event that the bar was raided by vice police. Clearly, managers on the premises have long served as buffers between owners and potential cases against employees of their bars.

B-girls continue to be depicted as criminals or criminal associates from the mid-1940s, through the 1950s. In March of 1950 we see the first nonfiction reports of B-girls as perpetrators of "knock-out drug" cases. These B-girls, sometimes working with accomplices such as club owners or taxi drivers, were accused of drugging men in the French Quarter. B-girls had shifted their attentions to civilians as victims. Such B-girls purportedly administered drugs through drinks and then robbed their impaired victims or forced them to write checks (*Times-Picayune*, March 30, 1950, 2). The period of rebuilding, between World War II and the beginning of the Cold War was a time when Americans were vulnerable to moral panics. These were symptomatic of wider panics—about economics, gender roles in the post-war economy, and social change. The early 1950s were a time period when queer sexualities (and here I would include sex workers) were scrutinized under the microscope of many Americans.

Locally, social turmoil in New Orleans during this time was tied to its economic status as a trading port, a city that relied on commercial business with very little manufacturing. Peirce Lewis (1976, 51) maps New Orleans' role as a link through which Latin American bananas and coffee passed to the interior of the United States. Changing relations with so-called "banana republics" threatened New Orleans' economy. Having long been entrenched in trade, postwar changes in transportation (with trucking challenging railways), the opening of Chicago to the Atlantic via the St. Lawrence Seaway, and new shipping technologies threatened New Orleans' economic dominance as America's leading port and gateway to the Midwest (Lewis 1976, 68). Further, the development of containerized shipping required less waterfront space, leaving the waterfront to return to its initial use as a space for leisure (Campanella, 2008, 232). Today's Riverwalk and Woldenberg Park are part of tourism-friendly development that became the focus of city planners between the 1950s and 1960s. It is in this context that the B-girl comes to represent a new sort of threat.

B-GIRLS AND TOURISM

By the mid-1960s, we find the first mention in *The Times-Picayune* of opposition between healthy tourist relations and B-drinkers. In 1968, the "Governor . . . asked for the states help in controlling 'vice, lewdness and B-girl rackets and outright fleecing of tourists" (Minor, Oct. 10, 1968, 14). Arguments appear in the paper between various city officials as to whether increased patrolling of nightclubs and crackdowns are good or bad for tourism—while no one likes to be fleeced, the presence of patrol cars could be interpreted as obtrusive to revelry.

Ambivalence about B-girls continues throughout the region over the next decade. One judge, who was criticized for refusing to give the mandatory minimum jail terms for B-drinking, stated that he viewed B-drinking as a lesser crime and that he would not send people to face the dismal conditions of Jefferson Parish jails for misdemeanors (Lee, Oct. 19, 1970, 6). Another article voicing concern for tourists as victims appears on August 13, 1979. The article focuses on the city of Bossier, Louisiana, where B-drinking had apparently slowed to a halt.

> The crackdown on the clubs came after a male tourist—loaded with a wad of money—complained he had lost $1,500 in cash and the diamond in his rings to one of the sweet-talking, fast-promising 'Hostesses' at one of the clubs. (Anderson, 1979, 12).

The story the paper tells concerning the wild success of the B-girls in Bossier is hard to believe. One wonders how these women managed to remove the gems so quickly, but apparently leave the rings. Still, the article ends with a string of interviews with men nostalgically recalling their own good times on "the strip."

The ambivalence of the B-girl character becomes increasingly clear in the mid-1980s. Several quasi-famous local women include stories of a B-girl past in their personal interviews, cementing the occupation in the lore of local legends. B-drinking is used as a playful contrast to such women's current legitimate success (Ball, April 20, 1986, C10). It is tempting to compare such stories to famed journalist Howard Jacobs' work, discussed in more detail below. Jacobs created several B-girl characters depicted as successful and clever university students such as "Sophomore Sue" and the "Carat Girl." Perhaps loosely based on real characters, to what degree Jacobs' work is reporting and to what degree fiction is here unclear.

Concern for tourists continues, alongside stories of famous visitors who are "good sports," buying B-drinks as part of their authentic New Orleans experience (Roesler, April 20, 1985, C1). Also in the early 1980s, the courts begin to interrogate the laws concerning B-drinking as lacking "clear and definite content" and of being gender discriminatory because the same behavior, legally, was permissible for men but not women given the language of the law. This issue was resolved in 1985 when a bill passed through the Louisiana House finally removed legal language specifying that B-drinkers were female (*Times-Picayune*, May 16, 1985, A30; June 29, 1985, A30; July 6, 1985, A14).

Film plots of the 1980s described by *Times-Picayune* writers include the term "B-girls," reflecting the common understanding of the term to the paper's readership. In 1981, Kim Novak's role in the 1955 release *The Man with the Golden Arm* (Preminger, 1955) is described in the paper as "the redemptive B-girl" (Times-Picayune, Dec. 6, 1981:7). In 1982, Marilyn Monroe's character is described as a "reluctant B-girl" in a television release of the 1956 film *Bus Stop* (Logan, 1956). Such descriptions suggest that the readership is familiar with the term and further evidences the general ambivalence of the role of the B-girl in the 1980s. In 1989, the nostalgic construction is complete with the film *Blaze*, based on the memoir of famed strip tease icon Blaze Starr published in 1974 (Shelton 1989; Starr and Perry, 1974). Starr's career, which began as a striptease headliner in Baltimore, positioned her forever in the history of Bourbon Street, not only because of her elaborate props such as the famous sofa-lounge she designed to produce smoke at the push of a button but also because of her notorious affair with Louisiana governor Earl Long.

A masterpiece of the literary South, John Kennedy Toole's Pulitzer Prize-winning *Confederacy of Dunces* (1980) includes the frustrated B-drinking Darlene. Aspiring to becoming an "exotic" at the Night of Joy on Bourbon Street, she is comedic in her stage failures. Written in the 1960s and released posthumously in 1980, the book rang true for locals. Today, a bronze statue of the main character Ignatius J. Reilly stands on Canal Street under the clock at the former site of the D. H. Holmes Department Store where the book begins.

By the 1990s, B-drinking had become embedded in the public discourse of New Orleans as a part of the French Quarter's tourism experience. Women who were successful B-drinkers at that time engaged in a variety of performances, especially nostalgia. Vintage styles like elbow-length gloves and cigarette holders took the performance off stage, and B-drinking provided guests with intimate interactions with Bourbon Street dancers. Despite the visibility of B-drinking, none of my 71 interview informants had ever been arrested for the practice. Arrests during this time period seem to occur only in combination with charges of "lewd dancing" or prostitution (*Times-Picayune*, June 20, 1989, B3; October 14, 1992, B8; June 22, 1993, B3).

Nostalgic interviews with veterans "ripped off by a B-drinker" during World War II continue to be presented as amusing anecdotes (Darby, Nov. 11, 1996, B1). Local ex-dancer, Allouette LeBlanc, on January 31, 1998 reminisced about her experience as a dancer in the 1950s alongside Blaze Starr, certainly in response to the movie about Ms. Starr's popularity. "You did have to B-drink—hustle bottles of champagne to the customers. That's how the clubs made their money" (Cuthbert, Jan. 31, 1998, E1) Like Marilyn

Monroe's character in *Bus Stop*, Ms. LeBlanc portrays herself as a reluctant B-girl who just wants to perform. Again on January 15, 2000, an article runs nostalgic stories of B-girls in the past, this time from the perspective of a local Bourbon Street personality, Frankie Ray Perilli, who made it to Hollywood. In the article "Frankie Speaking," Perilli discusses the famous Dunn case, a fatality that resulted from knock-out drugs in the 1950s. Although the article portrays the clubs as rough, that reality is buried in a remembrance of particular girls, funny stories about stealing business from other clubs, and a list of famous names that Frankie met working on the strip (Cuthbert, Jan. 15, 2000, E1).

The nostalgic re-creation of B-girls continued in 2012 with *The Clifton Monroe Chronicles*, a play meant to have "a playful naughtiness." Actors performing as a 1930s radio show cast in front of their "live studio audience" include a B-girl character (Mahne, Oct. 9, 2012; Oct. 12, 201,:A9). The B-girl, in this local theater, is depicted as a stock character of old New Orleans. She has officially earned her place in nostalgic storytelling.

HOWARD JACOBS: POWERFUL VOICES, POWERFUL MEN

Howard Jacobs, quoted above, was a society writer who created *The Times-Picayune* segment entitled "Remoulade," reporting lightheartedly on New Orleans life and culture for 25 years. Jacobs was nationally popular; his work appeared in *Reader's Digest*, *Newsweek*, and the *Saturday Evening Post*. Early in his career, Jacobs described the French Quarter as being "suggestive of a Hollywood movie set, with odd characters prowling about and mingling with the legitimate element" (Jacobs, Dec. 26, 1948, 4). A prototypical character he never neglects is the nightclub B-girl.

The quote that started this chapter, originally published in 1948, describes a typical B-girl encounter. In 1975, Jacobs reworks the quote into a new article. Returning to the "Berserk Reveler" as part of his discussion of the "function and background of the Sho-Bar," a club known at that time for its B-girls, he traces the history of the property bearing that name at the time. He includes the many names the business has borne, describing it as a "fixture on Bourbon since the 30s . . . distinguished by as many as 25 or 30 B-drinkers clad in evening gowns . . . " Jacobs tells us of various "legitimate acts" that the building has housed, contrasting such legitimacy to B-drinking (Jacobs, Sept. 15, 1975, 13). The answer to his initial question about the purpose of such clubs is simple: B-drinking.

The story has minor changes between 1948 and 1978. The hostess has become "voluptuous," the cost of the drink has doubled (from $1 to $2), and the client has become a tourist. With the shift from protecting soldiers to protecting tourists, the situation is less clear. The tourist in 1978 has become a "would-be Lothario," Lothario being a character in *Don Quixote* who is unscrupulous in his seduction of women. These changes are indicative of a shift toward increased ambivalence, equalizing the B-girl with a Lothario, neither of whom is innocent.

Jacobs' fictional works, such as *Charlie the Mole and Other Droll Souls* (1973), further contribute to the construction of the B-girl as one of New Orleans' many eccentric characters. In the chapter "Impromptu Floor Shows," he includes both the tale of the "berserk patron," and introduces the educated B-drinker Patti White, The Deciduous Pedagogue . . . "hoity toity for the shedding schoolmarm" (1973, 143). The book was reviewed in *Louisiana History: The Journal of the Louisiana Historical Association* by member of the U.S. House of Representatives and former newspaperman F. Edward Hebert. Hebert says:

> The people and the stories are authentic because Howard Jacobs knows or knew them. . . . The cast of characters depict an important part of New Orleans that often goes unnoticed. . . . As a newspaperman in New Orleans for 23 years before being elected to Congress, I met some of these folk heroes personally. Perhaps some of the names were different, but the roles played by these non-conformists are the same. It is like a play. The cast of characters may change, but the roles are always the same. (1974, 89)

The longest-serving member of the U.S. House of Representatives from Louisiana, Hebert represented the 1st Congressional District from 1941 until 1977. A homespun New Orleanian educated at Jesuit High School and Tulane University, before his election he was a front-page columnist and political editor of *The Times-Picayune*. Hebert's positive review of Jacobs' collection of French Quarter characters is telling of the government's laissez faire attitude concerning the antics of New Orleans characters, which are an expected feature of the city.

In "Remoulade" Jacobs narrated from the perspective of "the true Quarterite," claiming a voice of authenticity for a broad readership of New Orleans. He finishes the comedic piece, in which the B-girl is represented in one of her most common forms, the trickster.

> An hour later, when it dawned on the swain that she had played him false, his wrath was fearful to behold. Stalking into the floor show section in the rear, he proceeded to overturn a couple of tables, spilling drinks helter-skelter on the terrified occupants and bashing in the beezers of four patrons who sought to remonstrate with him . . .
>
> "My dear," later exclaimed a woman spectator who had occupied a table next to one of those overturned in the melee. "They have one act that is simply out of this world. It has nine people—a leading man who runs amuck and four men and four women stooges . . . " (Jacobs, Dec. 26, 1948, 4)

It is typical of Jacobs' humor that it is a woman who finds the B-girl's victim humorous—though not because she is aware of the potential for proximity to sex work. To her, they are all "stooges." The B-girl is a part of the satire. It is indicative of the New Orleanian character. To quote T. Harry Williams, the biographer of Huey Long (1969).

> Some have gone so far as to suggest that Louisiana is not really an American state, but a "banana republic,' a Latin enclave of immorality set down in a country of Anglo-Saxon righteousness. . . . Louisianians have always had a non-American attitude toward corruption. They have accepted it as a necessary part of political life, and they have even admired it when it is executed with style, and, above all, with a jest. (Quoted by Lewis 1976, 14)

The B-girl is a trickster, although who her victims might be depends upon the moral climate of victimhood and the shifting economy of this port city and the Gulf Coast more broadly. Although dangerous, she is also sassy and funny, an embedded person who knows the real French Quarter and can tell the reporter how it is. Finally, although these first two roles highlight her agency, she also crafts herself into an object which is a critical component of Bourbon Street as a site. A good sport who is navigating the touristscape of the French Quarter and Bourbon Street needs a B-girl, just as much as he needs to drink a Hurricane and don a set of (seasonal or not) Mardi Gras beads. Through the changing story of the B-girl we can see the relationship of Bourbon Street as a heritage tourism site to the changing moral landscape of the twentieth century as expressed through media discourses.

NOTES

1. As an anecdotal aside, I once took a young female friend from Tanzania on a tour of the French Quarter, and she asked about the strip clubs on Bourbon Street. I explained that men went to such clubs to drink and spend money to watch women dance with their breasts exposed. Her response was a puzzled "Why?" Here, beyond B-drinking, we can see that the entire concept of the gentlemen's club is a highly culturally constructed phenomenon.

2. Elected for four consecutive terms from 1904 until 1920 and a final term in 1925, Behrman's administration was responsible for considerable improvements in the city's sewer and drainage, streets, and public services. He held the position for longer than any other mayor in the city's history.

3. The structures on the original Iberville housing project have been razed since the 1990s. The original New Deal buildings existed until Hurricane Katrina when it closed temporarily due not only to damage but also to a general restructuring of public housing in the city. The site now contains the Bienville Basin Apartments, a mixed-income and mixed-use planned neighborhood.

4. Ms. Lewis reported her income from B-drinking in 1946 to be roughly $20 per night (*Times-Picayune*, Oct. 24, 1946:4). To contextualize this income I turned to a report co-written by the United States Department of Commerce and Bureau of the Census (Sawyer and Capt, 1948) which reports that the average income for nonfarm families in 1946 was about $3,000, and for nonfarm individuals not in families about $1,100. The average income for all urban and rural nonfarm families and individuals combined was $2,600. This means that at $20 per night

Ruth would have had to work 130 nights per year to earn an average income—or 150 nights to earn the average yearly income of a family. If Ruth were supporting only herself, for example living in a small boarding house, she would only need to work 55 nights to earn the average income for an individual living alone or with nonfamily members.

5. I use Carol Leigh's term "sex worker" to maintain my focus on a labor perspective. What all B-drinkers share is emotional labor included in many parts of sex work and in most cases implicit engagement in a wide range of occupations through a willingness to find "the right girl for that customer" (Interview, July 2014).

6. Peirce F. Lewis (1976) refers to Bourbon Street as an important component of the few blocks of the touristscape as most visitors experienced it at that time. Although as he woefully predicted, the touristscape has since expanded, Bourbon Street remains the icon within the icon, the attraction within the French Quarter that everyone visits. This means that, even for tourists who do not spend in the nightclubs, their existence pulls tourists to the city, and therefore is an important component of (particularly convention) tourism.

Chapter Three

Finding B-girls in the Ideoscape

Legally Defining the B-girl

The legislature played a pivotal role in the introduction and maintenance of the term "B-girl" over time. Important contributions to the construction of the B-girl as an imagined person can be found in documentation of courtroom discussions about B-drinking laws. Legislative documents written to prohibit the *strategy* of women earning an income soliciting bar customers for drinks cemented the notion of a *category of person*, the "B-girl," into the ideology of the region. The original laws describing B-drinking were persistently appealed on two grounds. The first was for being vague in description of what acts were prohibited. Secondly, the law was frequently appealed as discriminatory in not providing equal protection to men and women.

There is intersectionality between the law and its representation in media. Court cases and appeals were covered in papers like *The Times-Picayune* which were read by most New Orleanians. The media served as a vehicle to transport the term "B-girl" from the courtroom into quotidian discourse, into everyday speech. The introduction of the term to the broader public in this top down manner definitely played a role in the construction of B-girldom as it was perceived by the public. Certainly, from the perspective of the individual and his or her experiences, the legality of one's occupation and identity are tightly tied.

Careful reading of appeals to cases when the law was applied, including discussion in the courtroom about the law's phraseology, is helpful to a historical reconstruction of the categorical B-girl. The history of the practice lays the foundation for the contemporary practice. The term "B-girls" first appeared in the Louisiana legislature, under "Acts Prohibited" in establishments where alcohol is served (Revised Statute [R.S.] 26:90, 26:286). As of

2011, Revised Statutes to title 26 list B-drinkers at (R.S. 26:7). The B-girl question recurred in the Louisiana Supreme Court and is here traced through the second half of the twentieth century.

If B-girls have been invisible in academic texts, the same is not true in the foundational texts of Louisiana politics. Beyond the B-girls' introduction in city ordinances and the state legislature, appeals to convictions in the courtroom provided a place where the "B-girl" definition was further refined. Beyond talking about B-drinking, the courtroom is a place where laws are questioned. We find a meta-discourse of B-girldom—more argument about the definition of B-drinkers than about the activities that are prohibited. The history of employing B-drinkers can be learned by following this paper trail.

Far from a "forgotten law," as some of my interview participants suggested, plenty of discussion has occurred in the Louisiana Supreme Court concerning the potential lack of clarity and perceived gender discriminatory wording of the law. For decades, individuals appealed unfavorable verdicts following B-drinking arrests on the grounds that the law is poorly written or unconstitutional rather than simply on their own claims of guilt versus innocence. B-girls were not forgotten, but rather continued to be embedded in the courtroom dialogue half a decade after their origin in the law.

OF MINGLING, SOLICITATION, AND IMMORALITY: B-GIRLS' LEGAL LINEAGE

The regulations that concern B-drinking are found in both the city and state laws of Louisiana. New Orleans city laws are available as public record. (municode.com, at writing current as of July 17, 2014). The pertinent city laws are found in Part II of City Code, Chapter 10, which concerns "Alcoholic Beverages," and do not use the verbiage "B-girl" specifically but rather describe prohibited circumstances for licensed alcohol servers.

The applicable city laws are embedded in Article III, which deals with restrictions and regulations on laws on alcoholic beverages, under Division 5, "Other Prohibited Acts." They begin in Section 10-432 and run through Section 10-435 of the New Orleans, Louisiana Code of Ordinances. Such ordinances concern the rules one must follow to operate a place of business that sells alcoholic beverages, and Section 10-432 speaks broadly to the illegality of employing "persons to solicit business." Establishments selling alcohol for consumption on the premises may neither employ nor allow the presence of "any person, whether paid or unpaid, for the purpose of requesting, inducing, or soliciting any patron or person to purchase for himself or . . . any person any beverage or other product . . ." Section 10-433a says this applies to solicitors paid and unpaid, and 10-433b clarifies that employees such as the bartenders and waitresses that are central to this study cannot request drinks for anyone besides the "customer being served."

All strip clubs in New Orleans and the surrounding Louisiana parishes[1] served alcohol in the 1990s and continue to do so today. Certainly within the touristscape of the French Quarter, alcoholic beverages were a part of the anticipated "widgets" the consumer expected to find when they entered a strip club to watch the performances and interact with the entertainers. The construction of the tourism experience in New Orleans most definitely is steeped in a pressure to drink. Tourism-related photos, as well as artistic representations in the galleries that line Royal Street, commonly depict drinks and people drinking. Elaborate drinks like the Sazerac, Hurricane, and the more recent Hand Grenade have overridden the traditional Mint Julep, enjoyment of the South continues to be represented on a porch or balcony with glass in hand.

Additional relevant City Ordinances for a study of the strip clubs of the French Quarter are Section 10-434 and 10-435, which define the legal limitations on nudity of employees. Alcohol cannot be served to anyone

> unclothed or in such attire, costume, or clothing as to expose to view any portion of the female breast below the top of the areola or any portion of the pubic hair, anus, cleft of the buttocks, vulva, or genitals. (10-434a[1])

Additionally, no such person may legally "mingle with the patrons" (a[2]). Although this city ordinance does not reference the term "B-girl," it is clearly constructed to prevent the *activity* of B-drinking, particularly in burlesque clubs. The sections immediately preceding those on soliciting business prohibit lewd behavior and prostitution (Sec. 10, 430–431), suggesting these are related phenomena in the eyes of lawmakers. Before that is the prohibition to keep anaesthetics [*sic*] on the premises where alcohol is served (Sec. 10-429), perhaps in response to the alleged "knock-out drug" cases involving B-girls during the late 1940s and early 1950s.

Section 10 prohibits venues serving alcohol from featuring "acts which simulate" a long list of sexual acts, among them masturbation, oral copulation, and flagellation; all of these were common parts of the "floor show"[2] portions of a dancer's time on stage during my observations in the 1990s. Among my interview participants, although *none* had ever been arrested for B-drinking, one had been arrested based on the arresting officer's perception of her stage performance as violating this section of the law. During her floor show, she included removing, kissing, and mouthing her shoe in what was interpreted by the arresting officer as a simulation of oral sex using the high-spiked heel. Perhaps more hygienically, many women sucked candy in the 1990s in ways that certainly violated this law. The use of candy as a prop while between sets was common, although less common while on the stage.

The stage is a shared space, including shared poles grasped during potentially dangerous performances, and dancers all cooperate to keep it clean.

"Burlesque or striptease shows" are explicitly addressed by Section 10-435, which states that dancers are required to perform on a "stage or platform" at least 18 inches above the floor. In fact, the wording of this law suggests it applies to all shows and "exhibitions" in any establishment selling alcoholic beverages. This law, found in the same section of the city ordinances that forbids touching performers' "breasts, buttocks, anus, or genitals," is positioned in the law to address gentlemen's clubs, though they could undoubtedly limit any stage performance. The law was sporadically followed throughout the 1990s by women who performed table dances, private performances in VIP rooms that take place at an individual table during which eye contact and attention are on an individual client rather than from the stage. During times when "vice cops" were expected, table dances were performed while precariously balanced on high heels and standing on an overturned milk crate. A milk crate or similar box was widely believed to qualify as such a platform although they were often placed strategically between men's legs or not used at all.

Section 14 concerns the prohibition of obscenity as it relates to entertainment. Although I focus on B-drinking, this section is obviously relevant because striptease nightclubs are the most common settings for B-drinking. It is impossible to understand striptease in the French Quarter apart from B-drinking in the French Quarter. The jobs are coterminous, particularly when the customer is a tourist.

Section 14-14 forbids any person to "prepare, advertise, give, present or participate in any obscene, indecent, immoral or impure show or entertainment . . . " Based on the overall lack of arrests of women in the sample interviewed for this research, one can safely assume that community standards, as interpreted by the acting New Orleans police department throughout the 1990s, were met, and that the striptease and other erotic performances discussed by my informants were considered in line with local notions of decency.

APPEALS IN THE LOUISIANA SUPREME COURT

Documented appeals made by individuals who were convicted of crimes related to the soliciting of drinks by women in nightclubs allow us to see a long-term, historical commitment to the conservation of such laws. Appellate documents include information about why the convicted person(s) disagrees with the outcome of his case, as well as the logic that the Louisiana Supreme Court used to either overturn or reaffirm those decisions. Thus, in the records of appeals to the law, we find an ongoing conversation about what it is, exactly, that is prohibited. As a parallel to talk about B-girls in the media, the creation of B-girls in ideology can be found in the dialogue happening across the decades in the courtroom.

This section was written through a search of cases in which those individuals who were charged with violations related to B-drinking appealed their cases in the Supreme Court of Louisiana. Similar to the newspaper, the courts are a place where B-drinking is discussed in ways that reflect the ideology of the community. These voices are those of judges and other powerful men, both representing the State of Louisiana and defending B-girls, managers, and owners of such nightclubs. The gendered nature of the law is here not at all congruent with a contemporary discourse of protecting women from victimization, of preventing women from being coerced or potentially harmed by the practice. Rather, it shows a long history of characterizing B-girls as female predators who seduce and deceive, and a historically awkward insistence that the gendered wording of the law was rational.

Case 1: State of Louisiana v. Jolande Almerico (232 La. 847, 95 So.2d 334)

On April 1, 1957 Jolande Almerico appealed charges that had been brought against her by the State of Louisiana. Ms. Almerico held a retail dealer's permit for selling alcoholic beverages at the Manhattan Club, located at 1310 Canal Street. This neighborhood is no longer populated by clubs, which currently are centered on a thin strip as illustrated by the map of licensing in Figure 3.1. It is perhaps the case that this reflects the gradual expansion of B-drinking as a strategy from the former Storyville to surrounding neighborhoods. By the 1990s the practice mainly centered on the cluster of nightclubs along Bourbon Street, at least 7 blocks from the Manhattan Club's 1957 address. The address is, however, a mere block from the former legal red-light district of Storyville.

Figure 3.1 shows the blocks of the French Quarter that in 2014 were zoned for commercial businesses, ranging from tee-shirt shops on Decatur Street to the antique shops of Royal Street. The entertainment zoning area is home to the nightclubs of Bourbon Street, as well as other tourist attractions like the Preservation Hall jazz club. The map also shows the locations of establishments with live entertainment permits, which would include bands, DJs, and dancers. The map was created utilizing data packs obtained from the City of New Orleans website (nola.gov) and ArcGIS. Zoning areas outside of the French Quarter, and therefore outside of the scope of my research, were removed for clarification. The live entertainment permit data was updated in January of 2014. Evidence at the Manhattan Club was obtained beginning at 2:30 am on August 4, 1955 (the appeal is inconsistent, as in some places it reports late at night on August 3), when a police officer in

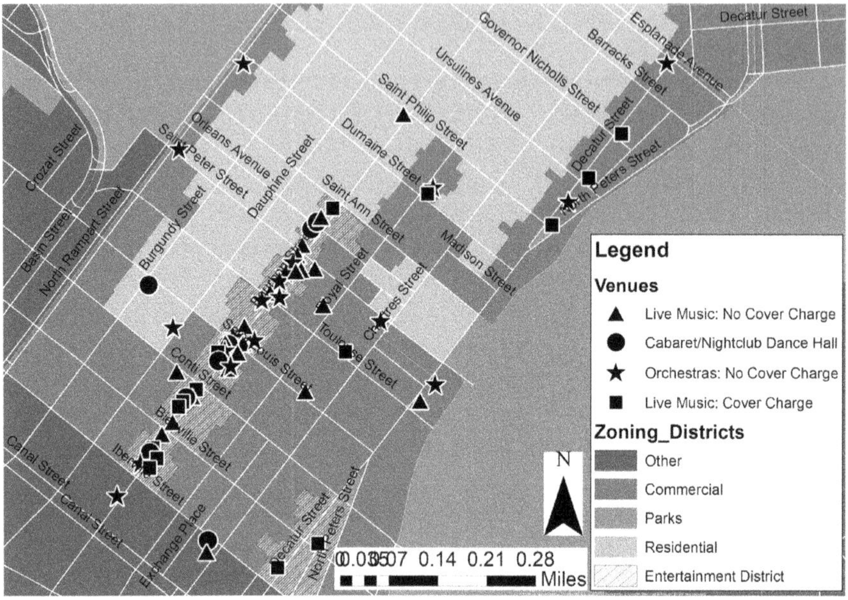

Figure 3.1. Map of Entertainment District and Permit Map. Created by Rachael Sebastian and Emily Jones

plain clothes was sent to the club. Mae Thomas, as the waitress, asked whether he preferred a table or booth and seated him. He ordered a beer and was joined at his table by Dolores Span, who solicited him to buy her a champagne cocktail, for which he paid $2.60.[3] Ms. Span suggested to the officer that another woman, Honey Lee, join them for a drink. Honey did join them, and after receiving her drink the two women suggested that it would be less expensive to buy them a bottle of champagne for $14.55. It was at this point that he identified himself as a police officer. Upon this revelation, Ms. Span fled to the back of the club, followed by the officer. While she successfully evaded him, Ms. Lee likewise fled, most likely to the front of the club.

Another plainclothesman entered the bar at "around the same time," and observed that the bar owner (Ms. Almerico) was present in the barroom. The appellate document reads: "The officer was *accosted* [emphasis mine] by Swannie Connors, who conducted him to a booth," and he ordered a beer. When the waitress returned she brought another woman, Ruth Johnson, who solicited him for a drink costing $1.30[4] while two additional plainclothesmen sat in a second booth nearby.

In the courtroom, Ms. Almerico challenged the case, arguing that the specific charges did not "disclose particular acts" through a "bill of particulars." In past cases, the defendant charged with such crimes was given such detailed indictment, but in this case the identities of the alleged B-drinkers were not divulged. Additionally, the state failed to prove the women involved had been paid, according to Ms. Almerico.

Ms. Almerico additionally contended that the evidence failed to identify those women as members of the classification "females, commonly known as B-Girls." The judge in this case found that it was not necessary to identify the women as such because "the statute sets out what constitutes the offense and merely refers to such females as being 'B Girls.'" That is to say, it does not matter whether the women's identities were established or whether they would self-identify or be identified by others using such a term. The judge here says that the term is simply a referent, which is clearly defined in the law. This theme, addressing the vague phrasing of the law, would continue to come up in the Louisiana Supreme Court into the 1980s.

The evidence listed in Ms. Almerico's case was the combination of the testimony of the police officers and three order pads that were found hidden in the flower boxes of the establishment. These order pads included two columns of numbers, a practice still in use in the 1990s, whereby the first column designates the B-drinker by some assigned or chosen number and the second indicates the amount she is owed for drinks solicited. In the case of the Manhattan Club, the judge acknowledged that the state must rely on circumstantial evidence, as direct evidence of payment to the women would be difficult to obtain. The judge found the law to be clear in "what constitutes the offense," and that the original sentence of a $305 fine and three months in Parish Prison (with an additional 60 days in prison for defaulting on the fine) was upheld.

It seems that this arrest was one of a series of raids that occurred in late 1954 and the first half 1955. In December of 1954, a special assistant to the city's attorney's office, Beuker F. Amann, along with a special police squad successfully filed petitions to revoke the liquor licenses of 7 establishments for serving minors and employing B-girls (*Times-Picayune*, December 16, 1954, p. 66 "Liquor Licenses May Be Revoked"). On January 5, 1955, an article ran in the *Times* entitled "Night Club Job told by Girl, 13" (page 7) in which a 13 year old from Arkansas testified that she had been employed "as a strip tease dancer and a 'B' drinker" by Nick Carno, operator of the Old Opera House at 601 Bourbon. Carno's crime was not only concerning the solicitation of drinks, but more generally the permitting of a person under 17 to loiter in an establishment selling alcoholic beverages as its primary business. At the time of the interview, the girl had been committed by juvenile authorities to the Southeast Louisiana State Hospital in Mandeville. Again on July 17, 1955, the *Times* reported that a vice detail raiding French Quarter nightclubs on two consecutive nights had led to 11 persons being held by police. On October 14, 1955, another article announced the arrest of a nightclub owner and four employees in a B-drinking case.

This string of arrests did not escape the attention of the society section of the paper—and on November 3, Jacobs' column "Remoulade" reported on a "gendarme-baffling device" reportedly borrowed from a similar affidavit that was being used in Boston. Meant to "confound the vice squad," women had reportedly begun producing a document to be signed by customers before being allowed to purchase drinks in "some Vieux Carre bistros specializing in unattached damsels." Described as an affidavit, the document was meant to allow the bar operator and B-drinker to use as evidence in court, as it stated that the signed has "not been solicited to buy the drink, and that he is doing so under no compulsion, inducement or prodding whatsoever." A municipal court judge is said to have refuted that such a document would be held as conclusive evidence, even if signed by vice squad men.

Case 2: State of Louisiana v. Molly Daye (243 La. 725, 146 So.2d 786)

On November 5, 1962, Ms. Molly Daye was convicted of permitting B-drinking and she appealed. She was charged on two counts of permitting B-drinking at the French Casino at 1428 Canal Street. The first charge was for permitting the solicitation of drinks of high-alcoholic content, and the second for permitting the solicitation of drinks of low-alcoholic content. In this case, the sentence was set aside because the law states that a violation can only occur by a "person holding a retail dealer's permit" or his/her employee or representative in both the statutes concerning establishments serving low and high (more than 6 percent by volume) content. In Ms. Daye's case, the judge determined that the court had failed to allege she was the holder of such a permit. Thus Ms. Daye's complaint that they could not prove the drinkers had been compensated, nor the type of drink served, was not reviewed by the court, because the indictment had failed to specify all circumstances necessary for the offense.

Ms. Daye's case contributes to our understanding of the court's view of B-drinking and its relation to alcohol. Based on public representation in newspapers that conflated stories about B-drinking arrests with arrests for employing minors during this same time period, one is tempted to conclude that such laws are meant to protect young women from alcohol and other purportedly related vices. In fact, in this case, Supreme Court Judge J. Hawthorne held that the alcoholic content of B-drinks was irrelevant in the eyes of the court. "In other words, the statutes are concerned with the solicitation of drinks, whatever the substance or kind of the drinks may be. If this were not so, the statutes could be circumvented by serving the 'B Girls' soft or non-alcoholic drinks." Thus it is not alcohol sales that truly constitute this crime, but the interaction between customers and female employees for which the woman (or girl) is waged that is problematic in the eyes of this judge. It may be significant that Ms. Daye's dismissed sentence of 60 days imprisonment in parish prison with 30 additional days for failure to pay the fine (still $305) was significantly shorter than the 90 day sentence, with 60 additional days for default to pay the fine, found in the previous case of Ms. Almerico. It may simply be that Supreme Court Judge Hawthorne was more sympathetic to such cases than Judge Ponder, who upheld the judgement against Ms. Almerico in 1957.

Case 3: Frances G. Burnette v. the Louisiana Board of Alcoholic Beverage Control and the Police Department of the City of New Orleans (Westlaw, 252 So.2d 346)

On August 26, 1971, the Alcoholic Beverage Control Board suspended the license of Frances Burnette for seven days for a B-drinking violation. Burnette appealed on the grounds that the law was unconstitutional because it was vague. In the resulting case, the Court of Appeal of Louisiana, Fourth Circuit, affirmed the initial decision. The court ruled against her, rejecting the claim that the law was unconstitutional for vagueness.

Ms. Burnette's claim rests in part on the discriminatory nature of the law. She claimed that the law denied equal protection "as to the appellant and alleged 'B Girls' as compared with other entertainers." Ms. Burnette's defense tried to point out the gendered nature of the law. The court found that she was not being denied equal protection, because in this case it was not the entertainers (specifically, the B-girls) to whom the law was being applied, but rather the licensee (Ms. Burnette, who happened in this case to be female). The gendered term "B-girls" refers to the *employees*, not the licensee, who under the law is not referred to in gendered terms. The law was found to be non-discriminatory in that the licensee, regardless of gender, is provided equal protection under the law. To quote,

> the phrase commonly known as "B Girls" is simply descriptive wording and is not a limitation on the statute, which *prohibits all Females* [emphasis mine] from soliciting patrons for drinks and accepting drinks from patrons for remuneration on the licensed premises. We find that the statute clearly and adequately describes the act or conduct which is prohibited.

Burnette's case involved two plainclothes police officers who entered a bar at 225 Bourbon Street, the Blue Angel,[5] and ordered two drinks at $2.55 each. When the waitress, Maria, returned, she was accompanied by a female named Sherry, who was alleged to have solicited the officer for drinks. The police testified that they did not remember what the women ordered, but that the cost was $3.50 for each of the women's drinks. A second round of drinks was solicited and paid for. When a third round of drinks, which were allegedly not ordered, was delivered to the table the men refused to pay and identified themselves as police officers.

Although again the court acknowledges that it is impossible to directly prove the women were paid for these drinks, they suggest that circumstantial evidence supports such a conclusion. In this case, the price differential in the drinks ordered by the police and the women was not alone considered sufficient evidence. However, combined with their solicitation by a dancer, the women's admittance of employment by the establishment, the very short time (15–25 minutes) of the entire episode, and the third set of drinks arriving without having been ordered led the judge to agree there was sufficient circumstantial evidence of the crime. The officers involved in the case did not have evidence that the women were actually being paid a commission to drink. The question, really was whether the "females were *not rendering their services gratuitously*[6] but were in violation of regulation prohibiting solicitation by "B Girls." [6] Intoxication Liquors 223 108.5)

This case is the first evidence of questioning the constitutionality of the gendered component of the law. Despite the court reiterating that the descriptive wording of the law applied to "all females," Ms. Burnette lost her appeal and the court upheld the delayed seven-day suspension of her liquor license, which began the next day, August 27 of 1971. She was additionally placed on probation for the rest of 1971. This was in response to her initial appeal with the Civil Court, which granted a restraining order against the Louisiana Board of Alcohol Beverage Control prohibiting their suspension of her liquor license beginning July 13, 1971, pending the review of the court.

Case 4: Jack Nichols v. the Louisiana Board of Alcoholic Beverage Control (257 So.2d 484)

This case was tried on January 31, 1972, in the Court of Appeal of Louisiana, First Circuit. It concerns a bar called the Trailer House Lounge in Morgan City, St. Mary Parish of Louisiana. Jack Nichols was the permittee for the establishment, as well as half owner. In this case, the Louisiana Board of Alcoholic Beverage Control was appealing the judgement of the trial court which had restored the liquor permit of Jack Nichols. Judge Landry did in this case reinstate the decision to revoke Nichols' State Liquor Permit Number 8146.

This case is based on a visit on December 16, 1970, at about 10:30 pm by an agent of the Louisiana Board of Alcoholic Beverage Control, Joseph Angelle. He entered the Trailer House Lounge as part of a "routine check" on the business. He reported that 10–12 people were present, and that two women he described as "barmaids" came from behind the bar straightaway to sit on either side of him. On their request, he ordered them each a $2 champagne cocktail as well as a drink for himself.

Both women asked for additional drinks and then attempted to solicit him to purchase them each a $25 bottle of champagne. He purchased neither. However, both women acknowledged to Agent Angelle that they had been employed by the lounge for "a few weeks." Angelle denies having observed either woman mixing drinks, wearing any sort of uniform, or keeping any kind of record of the number of drinks consumed (a form of evidence used in previous hearings above).

In this case, both the Almerico and Burnette cases were cited. The Almerico case is cited as setting precedent for the Burnette case. Judge Landry acknowledged the difficulty of proving receipt of payment on a commission basis for B-drinking. Instead, she suggests that in the Burnette case the indirect evidence of the price differential in the drinks, the fact that one woman was employed as a dancer as well as a waitress, the solicitation of drinks by females, and the rapidity of the event was admitted as direct evidence. In fact, a careful review of the previous cases shows that the price differential in the drinks was explicitly addressed in the Burnette case as *not constituting evidence* (contrary to this appellate document). Thus, although the court claims to adopt the interpretation of law in Burnette as its own, it does not do so with great care.

The result of this case was the reversal of the previous judgement to reinstate the license revoked by the board. As of April 26, 1971, Mr. Nichols' permit was revoked, and he was ordered to pay all costs of court proceedings.

Unique to this case is the decision that it was not necessary to prove the entertainer's remuneration was on a commission basis to constitute the crime of B-drinking. While Mr. Nichols stated he "did not exactly hire the females in question," he did admit that they were working in the lounge and that they were each paid a salary of $15 a shift, amounting to $87.50 weekly. To place this into context, according to the U.S. Department of Commerce the median income for American families (not individuals) in 1972 was $11,120—which amounts to roughly $213 per week for a 52-week work year. In this case, the court considered this a substantial amount of pay, perhaps considering the gender of those arrested and assumptions about men's roles as family breadwinners in the 1970s.

Based on the testimony in this case, although a lower court (the trial court) had found a fixed salary did not support a B-drinking charge, the State Supreme Court determined that the law did not expressly say B-drinking required a per drink payment. Instead, the court found that the combination of "a substantial daily rate of pay" and the solicitation of drinks was enough evidence to establish a B-drinking charge. It reports that to determine that this example "does not constitute a violation of the statute is to render meaningless the term 'remuneration in any other way.'"

Case 5: Gary DeFrances v. Edwin, Edwards, William Guste, and Henry N. Brown (480 F. Supp.1)

In May of 1977, bartender Gary DeFrances brought an action to the United States District Court, W. D. Louisiana, Shreveport Division, challenging that the B-drinking statute (La. R.S. 26:88(8)) was unconstitutional. Mr. DeFrances challenged the constitutionality of the law based on the Equal Protection Clause of the Fourteenth Amendment.

DeFrances bartended at the Black Knight Lounge on U.S. Highway 80 in the area known as the "Bossier Strip." Although he had not been arrested himself, several of the employees of the Black Knight Lounge and other bars on the strip had been arrested under the B-drinking statute. The bar, as described in the court's Finding of Fact, "employs women to dance in skimpy costumes and to serve drinks" and several women had been arrested for soliciting drinks for compensation under the B-drinking laws. The court record maintained once again that both men and women have been arrested on such violations.

The court acknowledged that the statute did prohibit conduct by women that was lawful for men. However, DeFrances' case was dismissed because the state had *not found* "B boys" in any establishment at the time of investigation. Since B boys were not known to the state, the difference in law was judged to be constitutional. Citing previous case law, the court maintained that statutory classifications, like this one that distinguishes between the sexes, were permitted if those classes of persons bore a "substantial relation to the object of the legislation" (Conclusions #3). The court deemed that the distinction between the sexes passed the "rational relationship test" in fulfilling state objectives. Thus, this case gives us some suggestion for the first time within the legislature of the objectives of such laws.

In the Findings of Fact, parts 4 and 5 of this case, the court record provides this as the "general pattern" of "the operation of B-drinking practice":

> 4. The operation of B-drinking practice follows a general pattern. A man enters a lounge. He is greeted by a costumed female who directs him to the bar or to a table. At a table she sits with him and asks him to buy her a drink. At the bar, she will sit with him and either she or the bartender will ask him to buy her a drink. The girl will remain with the patron until it is her turn to dance or wait tables, then she will leave to return later.
>
> 5. A key to the success of the operation is the sexually exciting atmosphere. Many of the B-girls insinuate or promise sexual favors to entice the patrons to continue buying them drinks. Very often the drinks contain little or no alcohol, as a successful B-girl may solicit between 40 and 60

drinks in an evening. The girl's drink costs substantially more than the man's.

6. On numerous occasions unsuspecting customers have found themselves short changed or their bills padded. Some violent disruptions have resulted from confrontations over such practices.

7. In addition, many of the women are treated almost like white slaves. Many of them are indebted to their employers for food, shelter, costumes, and the like. Some have been threatened with bodily harm if they did not perform properly. A few men have extracted sexual favors through duress.

The interpretation of what B-girls do leads the court to describe why it is a problem. B-girls are described as insinuating, promising, and enticing unsuspecting patrons to buy weak drinks. Such behavior, according to the court, leads to further shortchanging of customers; all of this leads to violent disruptions. This is similar to the idea of "secondary effects" that has been used in recent cases in arguments to close strip clubs or limit them through zoning laws.

This is also the first incidence, in tracing the lineage of the B-girl across court records, that any mention of the protection of the women is mentioned. This 1977 court record suggests that B-girls (but apparently not other employees of strip clubs) can be "almost like white slaves." This is the first time I find such a claim in the public records. This is an important shift in the discourse—within the same document we have the B-girls represented as both a potential criminal and a hapless victim. It is a subtle transition. The states objective, in this document, is preventing the commercial exploitation of both customer and B-girl. The B-girl's role is ambiguous, as both a perpetrator and a victim of extortion.

Having lost his case in the District Court in 1977, DeFrances took the case to the United States Court of Appeals in January of 1980 (609 F.2d 239). The circuit judges agreed with the previous judgement and affirmed the judgement.

Case 6: State of Louisiana v. Kimberly Bishop, Lynn Lewis, Tonya Baker, and Sue Gamble (356 So.2d 998)

This short case is obviously related to Case 5 above, as it concerns women who were accused of violating the law by acting as B-drinkers while employed by the Black Knight Lounge (where DeFrances was the bartender). Following the conviction of the women by the Twenty-Sixth District Court of the Parish of Bossier, the case was granted a *writ of certiorari*, allowing a higher court (Supreme Court of Louisiana) to review the decision.

In this case, the women won. The Supreme Court of Louisiana decided on March 17, 1978, that the defendants could not be convicted of employing B-girls, because they did not employ anyone but simply *acted* as B-girls. The women's council moved that the court quash the bills of information, as they had previously attempted unsuccessfully in the trial court. According to the court records, "the bills of information do not charge defendants with a crime known to the law of Louisiana," and thus the motion to quash was granted. This was the first case I can find where the court recognizes that acting as a B-girl is not illegal under the explicit terms of the law.

Case 7: State of Louisiana v. John Otto Cox, Jr., and Allen J. Breaux (352 So.2d 638)

This is an appeal made in 1977 by the state in the Supreme Court of Louisiana to a decision made by the East Baton Rouge Parish 19th Judicial District Court. The court had "quashed" (a legal term for when an action is declared invalid and therefore is annulled by the court) a bill of information related to charges of allowing B-drinking. The details of such bills of information serve as a written statement of accusation of a crime. Judge Donovan W. Parker had quashed the bill because he had agreed with the complaint that the statute in question was overly broad. This complaint had come up in previous attempts to avoid B-drinking charges in Louisiana courts. The Supreme Court continued to maintain the law was not overly vague, and that Judge Parker had been in error.

In this case, the owner (Cox) and bartender (Breaux) of Club Destination were charged with allowing B-drinking, specifically by Ginger Rau who was employed by the bar. Ms. Rau was accused of soliciting William Cox for drinks as a B-girl. William's relationship to owner, John Otto Cox, is unclear from the documents. The court carefully defends the completeness of the bill of information in question, including information about the licensing of the establishment and Ginger Rau's receipt of a commission; this inclusion was probably in response to earlier cases where bills were quashed for being incomplete.

Initially, the request to quash the bill was based on four separate grounds. The first was that the club had allegedly already been padlocked under order of the Alcoholic Beverage Control Board (ABC Board) for the same transaction. Thus, the defendant claimed this was a case of collateral estoppel and double jeopardy. Based on the same padlocking, the defendant also claimed that the statute violated the constitutional separation of legislative authority (the court) and administrative authority (the Alcoholic Beverage Control Board, which ostensibly padlocked the establishment).

The court found there was "no evidence to support defense counsel's assertion that such a padlocking occurred" (352 So.2d 638). The court points out that the double jeopardy would mean the contemplation of a second criminal prosecution, separate from any civil penalty to which they may (or may not, it seems) have been subjected. The court also points out that the defendant failed to raise a written motion for a plea of double jeopardy in line with court proceedings. Related, the court found that the defendants had no cause to raise the issue of improperly delegated authority because they were charged in court with violation of the Louisiana State Statute rather than a violation of the rules and regulations of the ABC Board. In short, the Supreme Court of Louisiana raises doubts that the ABC Board had padlocked the club, but even if they had, such actions were separate from the criminal case.

The judgement in this case also dismissed the idea that the statute was in violation of equal protection. The prosecution was against a person holding a permit rather than of a B-drinker or her patron, and the law equally punishes permit holders regardless of sex or gender. Whether an owner or bartender (the two defendants in the case) were male or female, the case would remain unchanged. Here we lead into the discussion, again, of overbreadth.

The court points out that the concept of overbreadth in the law is only applicable when prosecution is against a constitutionally protected right, such as freedom of expression as protected under the First Amendment. "The act of soliciting and accepting drinks from patrons in an establishment operating under a liquor permit is not conduct protected by federal or state constitutions."

In this case, the judge refers to two of the previously mentioned cases. First, the judge points out that the court had already determined that the word "drink," to refer to either alcoholic or nonalcoholic beverages, was not vague (Case 2 above, State v. Daye). Secondly, he points out that in Case 3 above (Burnette vs. Louisiana Board of Alcoholic Beverage Control) the court had already determined the phrase "commonly known as B girls" to not be vague but merely descriptive. The law, according to court documents, "simply" prohibited "allowing B girls to solicit patrons for alcoholic or other drinks . . . " According to the State of Louisiana, "The phrase 'commonly known as B girls' has a commonly accepted meaning, but even if it did not, the statute otherwise explains in detail the conduct which constitutes the offense . . . " Regardless, neither the owner nor bartender could challenge the statue on the grounds of equal protection, because neither of them were B-girls. The case was sent back to trial court for proceedings.

Case 8: State of Louisiana v. Erwin Corky, Doris Milazo, and Joseph Nuccio, State of Louisiana v. Daniel Umphlet and Rhonda Caldwell (458 So.2d 904)

In this combined case, heard on November 26, 1984, Louisiana Supreme Court Judge Lemmon rejected the previous decision of Criminal District Court of Orleans Parish, Judge Frank A. Marullo, Jr. Marullo had declared the "B-drinking statute" unconstitutional and quashed bills of information that charged the defendants with such violations. The Supreme Court of Louisiana once again held that the law was not ambiguous, as any potential ambiguity was in the phrase "B-girl," which was a descriptive term that did not diminish a clear understanding of prohibited conduct.

Supreme Court Judge Lemmon further held that the statute was a "legislative response to a serious problem of B-drinking by women in retail alcohol outlets." Although the defendants point out that similar behavior is not prohibited for men, the court argues that Louisiana state statutes may affect some groups differently so long as there is a rational relationship to a state objective. The objective here was to "prevent bar operators from hiring females for compensation to lure unsuspecting male customers into purchasing drinks for them." The judge cites the case of DeFrances (480 F. Supp.1—Case 5 above) had already satisfied the question of whether the distinction was justified. Indeed, the court document in this case suggests that the evidence from DeFrances and other cases provided a legislative history that this was a serious problem.

The court document reiterates the claim that those to whom the law is applied, those holding retail dealer's permit and employ B-girls, are a non-gendered group; the judge cites the previous findings in Cox (352 So.2d 638) as having already resolved this. In the same document, the court insists that the use of the term "B-girl" does not limit the law. Instead, "the statute clearly prohibits retail alcohol dealers and their employees from employing or permitting females to solicit and accept drinks from patrons for remuneration," whether they are commonly known as B-girls or not.

In this court document, a comparison is explicitly made between B-drinking and prostitution. The state cites the case of *State of Louisiana v. Devall* (302 So.2d 909), in which the court rejected the claim that the state's prostitution statute was unconstitutional because it was similarly gendered, prohibiting women from sexual intercourse for compensation without criminalizing the same behavior by men. In that case, Devall was cited as unable to provide evidence of similar conduct by men who "created a social problem of any importance" and thus failed to convince the court that it was unreasonable to focus on women. Although the defendant argued that "female impersonators" could create the same problems, the court stated that Equal Protection did not require a state to "choose between attacking every aspect of the problem or not attacking the problem at all."

In the end, Judge Lemmon reversed the decision of the trial court to quash the bills of information, and the cases were sent back to the trial courts for processing.

In the initial ruling by Judge Marullo, reported in *The Times-Picayune* on June 3, 1983, in an article entitled "Ban on 'b-drinking' Struck Down," he held that the ban was unconstitutional because of "an unequal application of gender." Lawyer and Loyola professor Basile J. Uddo represented Joseph Nuccio, the owner of the French Casino nightclub in his complaint. Uddo argued that "to suggest the phrase (B-girl) has any intelligible content, much less clear and definite, is sheer fantasy. In fact, no standard dictionaries even include the phrase." Further, the 1983 article suggests it was common knowledge that nightclubs employed men who impersonated females, who would therefore not be covered by the existing law. By the summer of 1985, the law was changed. According to an article published on May 16, 1985, in *The Times-Picayune*, the Louisiana House voted 89–0 to give legislative approval to a change in the law which would finally include men paid to solicit drinks in the new legal category of "B-drinkers."

NOTES

1. In Louisiana, the state is divided into parishes, which are a somewhat parallel concept to counties throughout the rest of the United States. This is but one lingering aspect of Louisiana's unique legal heritage, embedded in Napoleonic Code rather than British Common Law.

2. The term "floor show" was used by my informants for times in their stage act when they would, literally, dance from a position kneeling or laying on the floor. A floor show could begin, for example, by a girl going down into the splits and then rolling onto her side and then back. Not all dancers include a floor show. Floor shows are not particularly new to striptease. Blaze Starr's famous Couch Dance is similar to a floor show—except as a 1940s headliner Blaze was allowed to haul sofas on stage. Expectations of stage use in the 1990s did not allow for this.

3. \$2.60 in 1955 is roughly the economic equivalent of \$14.79 in 1995, during my observational research. This is amazingly consistent, as the cost of a "double" champagne cocktail in 1995 was around \$14.00.

4. Again, consistent with observations in 1995, this \$1.30 drink is most likely a "single" champagne cocktail. B-drinks purchased in the 1990s tended to come in single, double, or triple servings of champagne cocktails, with prices incrementally increased. For example, a bar may have \$8, \$16, or \$24 options, depending on the size of the drink.

5. This address is currently the site of Larry Flynt's Hustler Club, a strip club currently in operation but outside of the scope of my research.

6. Emphasis mine.

Chapter Four

The Cultural Geography of Power and B-drinking in the 1990s

In 1996, sitting in the VIP room of a club waiting for my friends to get dressed, I wrote notes to myself about the feeling of the panoptic within the boundaries of the place. Michel Foucault describes in *Discipline and Punish* (1975) the shift in the nineteenth century away from the control of society through physical punishment and its public display to institutions that rely upon the power of the gaze to control and discipline communities. On the back of a pad of blank guest checks, I noted the ways that employees responded to the placement of cameras. Experienced workers counted their money in areas that were on the edges of surveillance, such as a small space in the doorway between the VIP room and the main seating area near the stage. The ability of the manager to potentially be watching any interaction and the inability to gaze back—to know when one is being watched—effectively discipline the group to follow and even internalize and cooperate with systems of control.

The panoptic seemed well exemplified in the uses of space and the technologies of surveillance and power that I felt in that club. Like Foucault's machine, the space within the club is "a type of location of bodies in space, of distribution of individuals in relation to one another, of hierarchical organization, of disposition of centres and channels of power" (1975, 205). The behaviors of contract dancers and B-drinkers, employees behind cash registers, and customers with wandering hands were all controlled by surveillance cameras and mirrors. The potential that activities were under the gaze of the owner of the nightclub (or any manager, or even an unattended camera recording the space) affected interactions between workers who understood the system. I was made aware by both managers and B-girls that surveillance cameras and mirrors were sporadically used to control groups of women while they worked in clubs (also discussed by Egan 2004). It is, however, an imperfect panoptic, because those at whom the gaze is directed can see and communicate freely with each other.

Unlike strip clubs in geographically marginalized zones of northern cities, strip clubs on Bourbon Street operate very visibly as a feature of the tourism milieu on a major pedestrian thoroughfare. Using Richard Campanella's (2014, 269) survey of pedestrian traffic on Bourbon Street, the average number of visitors who passed in front of the Bourbon Street clubs under study was 40 every minute on weeknights and 95 per minute on Fridays and Saturdays.[1] A front-of-the-club manager is expected to keep his eye on the entire club, from the small sidewalk frontage to the back of the VIP room. Pedestrian traffic and the purposeful visibility of the businesses of Bourbon Street make observations about the spatial organization of deviant behaviors in strip clubs in other parts of the country, such as use of space to facilitate less-stigmatized drug use (Lavin 2014), less useful to understanding deviance on Bourbon Street. Although surveillance is ubiquitous, perhaps, to the contemporary strip club, in the French Quarter space around the club is more carefully policed for behavior *deemed deviant by the tourist* even if perhaps not by the employees.

The control of the space through surveillance included awareness from behind the closed door of the back office of the movement of police, known drug dealers, and street-based sex workers whose presence in front of or inside the property required management's attention. In a variety of ways workers tried to resist the panoptic gaze. Some dancers co-opted the use of mirrors to their own benefit (or amusement) in communicating with each other about customers. Knowing the limits of such surveillance also helped dancers to develop strategies to accept money from customers outside of the framework of B-drinks, minimizing the income they appeared to receive and thus the required tips they would be expected to pay out to coworkers.

This was one of many such field notes I took while waiting outside of the dressing room for a group of dancers, ready to go out for breakfast and a beer at one of several bars nearby that catered to service-industry workers. Most of the girls were trying to change into street clothes, wipe off some of their makeup, and get paid. Two young women were changing into street-appropriate evening wear and makeup for a "private show," surreptitiously arranged with customers that had closed the bar at 4 am in the VIP room. The doorman was dragging out the trash, an overflowing can of endless plastic cups, bar napkins, and empty glass champagne bottles. He couldn't leave the club until the last girl tipped him and left and the doors were locked, and, depending on the evening, he might also be expected to walk the manager (along with the night's cash earnings) to his/her car. There is a startling clarity to the use of space in the club when the house lights come on at 4:05 am, replacing the red and blue lights. The darkness that suggested intimacy in corners and the flashing multicolored stage lights disappear, and the brightness, like stepping into sunlight, changes perspectives of the space. I suddenly noticed the discarded cigarettes and dropped napkins in grungy corners, awaiting the cleaning company that came every morning sometime past 6 am.

From the vantage point of a sticky leather booth in the back corner of the VIP room I can see myself in the mirrors on all three of the walled sides of the open, bright room. The air conditioner groans as it turns off. It's a relief when the air stops circulating. The club is kept as cold as possible with an open door in the summer months, a strategy that emboldens passing potential customers to step inside to stay at least long enough for a hostess, waitress, or B-drinker to engage them in conversation. The temperature is set to entice men in jackets and ties rather than tee-shirts. For the employees, especially dancers sweaty after a set of three songs, the place was often too cold. The individual's phenomenological experience of temperature in the front stage of tourism interactions was hierarchized in favor of guests. Emotional labor often included a woman disguising her perception of the temperature. Some well-established performers invested in costumes that included additional layers, like wrap style gowns, to wear while B-drinking. However, costumes are expensive, and personal belongings are difficult to keep secure. Part of the embodied experience of B-drinking in the summer months on the Gulf Coast is being cold.

Sitting in the VIP area and looking out through the mirrors on the opposite wall of the adjoining space, I can see a reflection of the bartender Stephanos counting money in the cash register tray and drinking Hennessey. Stephanos has worked for a long time as a bartender, and he brings his own liquor to the club to prevent arguments with the owner about his steady drinking through every shift. The use of mirrors and lighting in The Den are an example of how space in the strip club is used as a part of the "disciplinary institution" (Foucault 1975) that organizes the behavior of employees, who know that they are intermittently watched by management. The workers in such clubs are Foucauldian "docile bodies," productive and obedient in a hierarchy of surveillance and power, who perceive this gaze as normal and generally interpret it as necessary for security. Stephanos has adjusted his drinking and economic practices to working under surveillance, reflecting Foucault's "link between an increased aptitude and an increased domination" (1975, 138).

The lights and mirrors also serve to direct the gaze of the tourist walking into certain parts of the clubs, and they suggest different levels of perceived intimacy available to visitors to each space. The darker, more recessed areas like where I took my notes are less clearly seen from the entrance, but the differences in lighting and furniture make them intriguing. Under the colored lights, the flamboyant red-velvet chairs along the now-empty stage are the focal point of activity from the doorway, and beaded curtains visually disrupted the gaze into the VIP areas. With the house lights on, places where sharp-heeled shoes and cigarette burns have already carelessly (or as an act of defiance) pierced the ostentatious red material become obvious. The black vinyl chairs in seating farther from the stage pop into view.

The girls are coming out of the dressing room and going to the bar for their pay, which is being paid directly from the cash register in a protected corner. There is a silent struggle going on over the division of the evening's income between the entertainers and the bar staff, rarely verbalized and the subject of open conflict but always watched carefully. Stephanos is behind the bar paying women for their work for the night as they come out of the dressing room. It's unusual. After most shifts the bartender takes the cash tray back to the office to be counted, and the manager pays the women from the desk.

There is a bit of chaos, as women flock to hand Stephanos handfuls of single dollar bills received as tips in exchange for twenties while he is trying to pay other employees. The bar rarely needs to buy single dollar bills as change for the register from the bank; the same dollar bills can cycle from the bartender to customers as change, to the tipped employees, and then back into the register in exchange for larger bills. Nobody wants to pay their electricity bill in single dollar bills. The bartender knows that if he needs to buy more singles than usual, someone is most likely hiding income by keeping small bills, reducing the amount they must pay out to other employees like the doorman.

Women who have had successful nights often ask to be paid in hundred-dollar bills. Many don't have access to banking accounts, and some choose to manage their expenses mainly in cash. Large bills are easier to save than anything smaller than a twenty. Still, some women avoid changing out their bills at work. In the skirmish to get paid, make change, and get out the door, differences in income between workers over the shift can come to other employees' attention and cause fights. Existing animosities between contract employees competing for business would be initiated or agitated when someone made a conspicuous display of changing out larger numbers of bills than other women at work.

Changing bills (usually after being paid in the office by a non-tipped manager) allows the bartender and doorman to have a better idea about each woman's actual income for the night. The combined share of tips owed from each woman was prescribed at between 5 and 20 percent of her combined income from any hourly wage, stage tips, B-drink payments, and tips paid in the VIP room, depending on whether she was a dancer, a waitress, or just B-drinking. This was true in two of the three bars I observed—in the third the doorman but not the bartender was tipped. Thus, everyone employed has an economic incentive to watch what is going on throughout the club.

Workers who conceals their total income can adjust their actual tip according to their own perception of service provided by the club employees, rather than being restricted to the tip amount prescribed by the club's management. Contract entertainers in the club who concealed their incomes have hard power over employees like bartenders and bouncers, who can be driven from their jobs by workers who refuse to "tip out" to them. The nightly pay of doormen could fall well below the minimum wage if they were rejected by the entertainers. Being fairly paid out of tips as owed easily doubled or even tripled the doorman/bouncer's nightly pay, which was paid in cash at the end of each shift. B-girls could also conspire with customers for larger tips rather than more drinks, especially within the VIP lounge, reducing the amount of alcohol sales made by the club overall. As B-drinkers engage in emotional labor with their clients, so too must successful management woo the contract workers themselves, flattering and joking with the dancers for their own tips. The emotional labor of bartenders, doormen, and even management is performed for the contract entertainers as an income-earning strategy, and charisma is a job requirement.

Stephanos was a favorite of the women and was often seen drinking with them in after-hours clubs. As a bartender, it could cause a conflict of interest for him to pay out the club's wages, but he rarely had conflicts with the entertainers about tips as he was well liked. When the owner computed wages earned and paid the entertainers, most bartenders had a more difficult time collecting their expected tip-out. The owner of The Den ignored when doormen like Mark made complaints about being under tipped. Rather than fight with the dancers and B-girls who were harder to replace, he dismissed the short changing of doormen as merited because of their lack of masculinity and power over women, therefore beyond the control of the office. On this night, instead of loitering in a disorderly line in the hallway outside the office, the women were sitting around the bar as they came out of the dressing room, watching each other and chatting. Mark was watching like a hawk.

Paying the girls from behind the bar gives the impression that the club owner is still in the back office but does not want to come out or be disturbed. He may be asleep. He may or may not be there—he is also a careful performer. Sometimes he arrives during a shift, dressed in black jeans and a silk shirt, and strides through the club to his office only to exit silently through the backdoor for hours or for the entire shift. Sometimes he arrives through the backdoor, and a shift of dancers and waitstaff don't realize he is in the office until he comes out, angry about something happening in the VIP or dressing room that he has seen on a security camera. Sometimes he sits with different B-girls for hours in a booth, giving the impression to potential customers that well-dressed men drink in his club.

After the last girl is paid, I watch Stephanos stroll out from behind the bar, carrying the tray of cash from the register the length of the club to the office. The building is long and narrow, like many other buildings in the French Quarter. You could draw a straight line through the front door; along the red carpet of the main walk space between tables; past the low-backed, red-velvet lounge chairs that surround the stage, parallel to the seating along the wall opposite the stage; and into the office. Two of the three clubs where I spent time are organized in this same way, although The Den is the nicest, with some paintings by local artists on walls in dark corners and generally untorn upholstery. Stephanos silently passes the opening to the VIP room, nodding slightly to me and disappears out of sight as he enters the mirrorless hallway that leads to the office.

The VIP room where I sit juts out toward the Canal Street side of the building, tucked behind the storefront of the business next door. It is not visible from the doorway on the street. I can't see out the front door to see if the doorman has turned off the lights in the club's neon sign, although I am sure the door is locked because of Stephanos' attitude. The sign is a source of recent gossip, because legal restrictions won't allow for a newer, larger sign given the size of the business' storefront. When the owner is bored he sits at the table closest to the dressing room with female employees. If there are few customers, girls sit with him and chat, and recently he has drunkenly, repeatedly complained about the sign.

"BACK STAGE" OF THE TOURISTSCAPE

From the corner of the VIP room I can see most of the bar and stage, directly or through the multiple mirrored surfaces. I can't see down the hallway to the office, which narrows as you pass the dressing room and bathrooms, spaces usually filled with dancers, B-girls, and waitresses changing or adjusting their makeup and costumes. The dressing-room door is always closed, and even if the door is open the tables and mirrors where the dancers are changing out of their costumes and into their street clothes is out of sight, around a mirrorless corner. Farther down the hall, past the dressing room and bathroom, is the office. This dark area is mostly for employees, except for when a guest uses the bathrooms. It is a highly gendered space—without invitation, men don't enter the ladies room or dressing room and women don't enter the men's room or the office. This is true for employees, customers, and even the police, who announced their entry beforehand the only time I saw them go into the secreted area of the ladies dressing room. On the three occasions I saw the owner of The Den enter the dressing room, he first knocked and sarcastically asked if he could come in. Dean MacCannell (1976, 101) suggests we think of Erving Goffman's ideas of the front and back stage as "ideal poles of touristic experience." Tourists desire to see what is authentic but never get to the farthest stages of the true "back stage," stage 6 in

MacCannell's typology. While the VIP room is within MacCannell's stage 3, a front region of the bar organized to appear to be a back region, the dressing rooms were this inaccessible stage 6.

The dressing rooms of all three strip clubs included here were distinct places in comparison to the front of each club where customers could meet. The office in which I began this book was more nicely furnished than any of the three dressing rooms of the clubs where I ultimately made observations. Dressing rooms were ubiquitously dirty, small, and poorly furnished.

In one club, an unfurnished second floor was used as a dressing room for the dancers and waitresses who needed to change into their club clothing and stage makeup. Duffel bags of clothing were on the floor here and there, and women complained there was no security for their belongings. There were only a few chairs, and women would struggle with shoe buckles sitting on stained and matted carpet. There was a bathroom and one mirror above the hand sink where a crowd brandishing mascara and makeup brushes could always be found. Women had to come downstairs in their stage clothing and enter the stage from the floor of the club. This dressing room was the source of complaints from the workers, but such complaints fell on deaf ears.

The Den had a more secure space. The dressing room had lockers for which contract employees could bring their own locks. A few women who worked on a regular basis could keep their costumes locked in their lockers between shifts, although an absence of more than a week usually resulted in the lock being cut off for another employee. Clothing tossed from lockers, however gently, was picked through by other performers for the choicest items in the days following such an eviction. The remaining belongings—panties, torn stockings, and oddly sized shoes—would end up in a pile on the floor from which indigent women could piece together costumes when they first started work.

When I look at images that I have in my collection, or that women who worked in the clubs share with me, the dressing room is by far the most common setting depicted. In the club from which I have coded the most photos, the previous beige paint shows through the visibly flaking-away blue paint, and exposed wires jut from a hole in the wall or dangle from the ceiling. Photos of smiling young women sharing drinks, hugging, and laughing are mostly framed by walls spotted in graffiti, graffiti art, bumper stickers, and smudges of makeup rather than the red-velvet chairs in the front of the club. In one photo, one of the girls is in full costume, orientated back in a seductive pose with one high-heeled foot on the makeup table; the chair is broken, missing a front leg, used less as a chair and more as someplace to perch while reapplying makeup. The dressing rooms of such clubs tend to be furnished with discarded furniture from the latest remodel of the club floor.

A social narrative about working on Bourbon Street could be found in the photos (all taken by employees themselves, mostly in the 1990s) that women shared with me over the course of my research. These 122 photos can serve as a cognitive map of sorts, reflecting the areas of the bar that are most relevant to the memories shared by group members. Roughly 47 percent were of women, posing alone or in groups, in dressing rooms. Of the photos shared with me, 26 percent were taken inside public seating areas of the clubs, such as at a table or at the bar itself. Only 15 percent of the photos were of girls on stage, and 12 percent were images of women taken on the street in front of the bar (figure 4.1). I included photographs taken in the doorways of clubs in this last percentage. None of the photos included customers.

This dressing room in The Den was elongated but narrow, with a doorway that opened directly onto the stage. Along half of one wall at hip-level was a wall-mounted, elongated wooden table, usually broken up into areas territorially claimed by the various large makeup boxes of regular performers. A mirror ran the length of this makeup table, and women often jostled for space. Whether people were starting a shift, sharing a break, or trying to get dressed and go home, the dressing room was often crowded, noisy, and filled with cigarette and marijuana smoke. A posted list of vague club rules, including not "showing your pussy on stage," was a seemingly semi-permanent

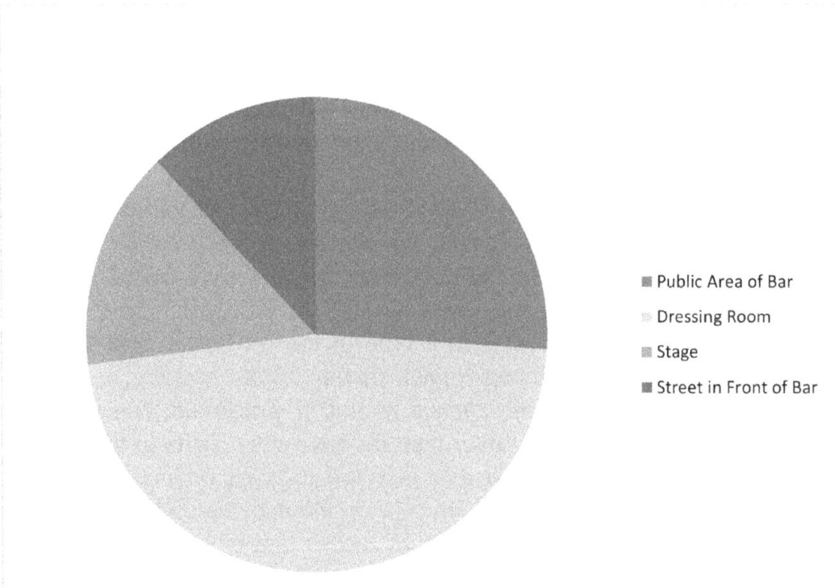

Figure 4.1. Photos Taken by B-drinkers Working as B-girls on Bourbon Street—Sorted by Setting. Created by author.

fixture, addressed at the top in large letters to the employees as "*Putas*" (meaning in Spanish and Portuguese whores, or perhaps more kindly sluts).

The roof leaked in the back of the dressing room for several years, and the employees who wanted to eat a meal during their shift shoved makeup aside temporarily to make space for their food. In one photo, three young women laugh and point up to a section of the ceiling that has completely collapsed in, revealing a tangle of electrical cords. The food, water, and nesting material (discarded clothing) was the perfect combination to attract additional residents to the dressing-room space. I will never forget the commotion in the dressing room on the evening that a large French Quarter rat ran across the crowded dressing-room floor, abandoning the remainders of a take-out salad to hide behind the lockers. One of the girls hopped over, still in her Doc Martin boots from the street, and slammed the locker against the wall to kill the rat. It was only after two weeks, when the smell of the dead animal could no longer be denied, that the rat was finally removed. For those two weeks the dead rat shared this space where the women dressed, applied stage make-up, took breaks, and ate their meals.

The difference in the attention given to the physical environment of employees and customers in any business varies and is to be expected. Additionally, labor laws that protect most employees do not apply to contract workers like those who labor on Bourbon Street. However, I don't think most of the women knew this or thought about the situation in this way. Rather, the physical environment of the space, like the use of the word *putas* on the sign concerning club rules, tacitly communicated messages about the value of the women who changed there. The back-stage setting of the dressing room was created as a part of reminding the contract workers of their role in the power hierarchy of the club. It is in the interests of the club that they internalize this message, which contributes to the internalization of stigmatization they face outside of the club.

This true "back stage" of the club serves an important role in what Pierre Bourdieu calls a social field (1986). In his theory of practice, a social field is a setting of interactions where economic, social, and cultural capital are mobilized and define individuals' varying positioning within the system, with the value of different sorts of capital being field specific (Bourdieu and Wacquant 1992). In the context of the strip club as a social field, control over economic capital is in perpetual flux. Those performers who are earning the high incomes for the club increase their own individual agency within the club, and a power struggle concerning management-set rules about tipping other employees can lead to the potential loss of a valued worker. Women who are successful at B-drinking also acquire embodied forms of cultural capital, Bourdieu's *forme incorporée.* Even if they do not have institutionalized forms of cultural capital such as diplomas, it is a part of their job to make conversation with high-end clients who do. The acquisition of mannerisms, ways of speaking, tastes, and other dispositions are directly transformable into economic capital. As a part of the effort to continually re-create a social field controlled by management, I suggest the dressing room is kept in stark contrast to the more comfortable public areas of such clubs. The dressing room, as well as clubs where the ladies' bathroom stalls no longer have doors, is a constant (re)creation and reminder of the overall symbolic capital of workers who must occupy such spaces relative to management within the club environment. Management can better control the club as a field when contract performers have relatively lower positioning in terms of their combined economic and cultural capital.

The booth-benches of the VIP room are part of the "front stage" of the tourist performance. The ashtrays are not just emptied but wiped clean by circulating waitresses. The high-backed wooden benches have an odd platform-like footrest along the wall under the seats, which B-girls use to suggestively straddle men without necessarily touching them. It is reminiscent of the kneelers in Catholic churches. For a short person like myself, this detail is noticeable—I can put my feet down.

The small moveable tables immediately in front of the booths each contain a crystal ashtray and can hold perhaps 4 glasses, although the space was built and designed for one customer and one B-girl. The tables can quickly be moved off to the side to accommodate a "dance box," a sort of tiny stage; tables can't accommodate bottles, which are delivered in a silver champagne bucket with a stand. There are no chairs facing the tables, only the bench along the wall on one side. This allows a VIP party an unobstructed view of the stage show, unless his view is temporarily blocked during a private dance by the B-girl or strategically thwarted by a waitress covering for the behavior of the B-drinker who is dumping half of her champagne into the ice bucket.

One summer The Den was requiring VIP workers to follow the law that obligated all live performances to occur on an elevated platform. The wooden dance boxes were stacked in a corner and disappeared a few months later. I hadn't seen the boxes in previous years. When I asked about them the club owner silently glared in anger and feigned ignorance, so I ceased pursuing a question that clearly was not my business. It was quite impressive to have seen my friends on six-inch, platform-heeled shoes balance gracefully on these boxes during private dances—Shadow, who is not a classic beauty but has training in ballet, took the opportunity to show off her elevated pirouettes. The boxes had also created a competition between dancers in the myriad ways and degrees of camp with which they could perform "falling off the box" into various "accidentally" compromising positions while in the VIP room doing parallel private dances. Perhaps most impressive of all was the seeming suspension of disbelief of tourists[2] as customers in witnessing the campy performances of drunkenness, reluctant willingness, and accidental touching with this maximum use of props. Many a fall had been soothed by the purchase of another bottle of champagne, adding $40 and a tip to the paycheck of the fallen.

Against the wall to my left, the leather seat back disappears into eternity in the mirror. To my right are the other seats along the back wall. Each booth-sized bench ends in a thin wooden separation that juts out about a foot from the wall, symbolically separating individual spaces for up to five couples to engage in discreet, intimate encounters for about $150 per half hour. I have seen these booths packed during conventions and empty during football games, when large groups of men who were less interested in private interactions filled the front of the club and surrounded the stage and the bar. Crowds of that size increase the number of single-dollar tips received by the stage dancer, but rarely produce the higher-end incomes from B-drinking. B-drinking requires a lower population density and more of a feeling of privacy. Thus, a crowded bar of festival tourists buying five-dollar beers can generate much less income than a moderately filled bar of business travelers with larger individual budgets. Larger crowds also meant more potential for conflict between tired performers and slow-spending customers.

If there were three or fewer VIP shows going on, the girls would space themselves out across the bank of five couches. Sometimes these benches could seat three cozily, as when a customer purchased bottles for two women. Sometimes men who came in together would be enticed by women who worked together to have a "double date" sort of VIP, which made it easier for the dancers to avoid being touched by the customer. Instead, they could verbally draw out any fantasies in conversation with their customers while physically engaged mainly with each other.

The Den, the club where I had the strongest community ties, has been a strip club for more than 70 years. Sometime between the club's height of fame in the 1940s and my time of observation in the 1990s, the upstairs balcony of the historic building had been closed to audience members. It was never reopened to customers, over more than a decade of my visits to Bourbon Street. This was important to my understanding of the history of the place later, as accounts of women who appeared in the same building decades before suggested much larger audiences than would be possible in the 1990s. The disinterest in reinvesting in the upstairs seating area was perhaps a reflection in a change in how business was done, or it was simply too expensive of an investment for the businessman renting the property during my observations.

In the early 1990s, a transgender performer named Ta'Mya was given the space just at the top of the stairs for her private dressing room. She had worked as a dancer for more than a decade, with many years spent at this club. When she invited me up (once), I could see that some bar supplies were also stored upstairs, but otherwise the place was filled with discarded furnishings and dusty artifacts of busier days. Ta'Mya's personal dressing space included a vanity with a lighted mirror and a small chair. Her mirror was lined with cutouts and photographs of famous striptease dancers that had passed through the buildings on this block of Bourbon Street during the 1940s and 1950s, as well as photos of girls that she had danced alongside since the 1970s. Her gorgeous sequined and feathered handmade costumes seemed out of place hanging on nails in the wall. She sewed them all herself, just as Blaze Starr claims in her authorized biography to have done. She has since passed away is deeply missed and was often mentioned in interviews and casual conversations with the community of dancers that made up my sample of participants for interviews from 2012 to 2016. At the time of her death, she was working as a "house mom," mainly helping women with their costumes and scheduling the dance list, but still earning an income as a talented performer in the emotional labor of B-drinking.

Ta'Mya was an interesting part of my sample because she was deeply integrated into the community of women working at that club. Although Blaze Starr remarks in her official autobiography that cross-dressing did not occur on the stages of Bourbon Street in her time, it did in the 1990s. In fact, Ta'Mya was integrated so well into the community that she was one of the few dancers who remained in the business past her dancing career, when she became known as "Mama" to dancers. Ta'Mya was not engaging in a drag-queen performance; she was a transgender person whose gender expression was feminine across all aspects of her life. She was treated no differently than other dancers in the 1990s and had a long and very successful career.

The club backdoor led to a courtyard behind the building. The narrow space between buildings was paved to create a walkway to the front of the building, with a locked metal door that opened to the street. From the street front, you would not know that this door was a part of the building at all. Before it was a tourist site, Bourbon Street was a residential street, and the big building would have had servants' quarters in the rooms surrounding the courtyard. By the 1990s the tiny apartments were set up as efficiencies which surround a courtyard, with bicycles chained up where gardens would have been.

Most clubs I visited in the 1990s included a VIP room (or Champagne Room) like The Den's back corner, or they created one by visual demarcation such as a drape. This is the most private spot possible for a customer to enter within the business. The center of attention for most of the audience is the stage, where girls take turns performing three-song sets. Every club's main stage had at least one pole, as pole dancing had become an important part of exotic dance by the 1990s. Women chose their own music, and the first song was selected in part for its strong introduction as the dancer would enter the stage during the first few bars, usually dancing in full costume. Parts of the costume could be shed for the second song, and the third song might be performed in as little as a wide thong panty and boots or shoes. In the early 1990s women covered their nipples with glued-on "pasties," which ranged from substantial disks of sequins with tassels to a thin layer of latex, but this practice disappeared over the course of my research. Many dancers used props, such as sheer dusters (a style of long, loose open garment originating in the fashion of the 1920s), feather boas, and corsets to emphasize or cover as they preferred throughout the act. Even transgender performers like Ta'Mya could dance in undergarments that exposed their derriere but covered their pubic area by carefully tucking their penises back and skillfully using props.

Men who meet dancers on the stage or on the club floor between their dancing sets purchase bottles of champagne to gain access to the VIP room, rather than sitting in the main area of the nightclub. In the 1990s, bottles ranged from $75 for a split, a small bottle of about 6 ounces, to a $1,500 magnum, a 1.5 liter bottle containing two regular bottles by volume. The most common bottle purchase was a standard-sized bottle of a sparkling wine that would sell in a liquor store for between $8 and $20 dollars. In The Den, this bottle cost the customer $150, with a $40 payment made to the entertainer at the end of the shift and the remaining $110 going to the club. This bottle, like a drink accepted by a woman from a stranger in any nightclub, represents her agreement to interact more personally with the customer. In the VIP room, girls would sell private dances, drape their legs over a customer, and engage in other private transactions. Because of the high cost of the VIP room, tourists who were engaging in holiday spending patterns were those most likely to buy bottles. Locals, who were in for a few beers and company, were more likely to be found sitting at the bar in the front of the building. A handful of such customers continue to be members of the social network today, mostly those who developed relationships with specific dancers outside the club as boyfriends or husbands.

A foul-breathed nightclub manager first demonstrated to me how he could see this back VIP corner from a chair at the very front of the club, sitting adjacent to the doorway, by using the images of images within the clubs mirrors. Laughing, he explained that between this and the use of cameras, privacy is an illusion only for the customer. Most employees were aware their behavior might be being recorded by security cameras. None of the employees I interviewed felt it was *their* security that was being protected using such cameras, but rather the interests and security of the club. I was once shown an image of myself in the club by an angry manager who hoped to gain something by this proof that I had been inside a strip club. Such use of photos to control women's behavior is typical of management styles I observed across the 1990s.

Design features within the public areas of the club were calculated to make customers feel they were in a private, intimate space while still allowing managers and staff to be aware of everything at any given moment. In one club, a small room-sized enclosure blocked off part of the VIP room from sight with a partial wall. The room contained a loveseat, a recliner, and a cocktail table for drinks. This most private space was walled off on three sides with wooden bric-a-brac meant to provide the customer a feeling of privacy while allowing any manager or waitress walking by a few feet away an unobstructed view. The same social contract that separates conversations by booth in any restaurant lends to the illusion of privacy within the VIP area. The reality is one of the panoptic; long-term employees know that they may be under surveillance at any time by other employees or cameras. In fact, women often depend on such surveillance when dealing with aggressive customers, as a passing waitress can disrupt the implied privacy and assist her coworker.

One club I visited took surveillance to the extreme of having torn the doors off of the stalls of the ladies' room toilets. To use the toilet in privacy, girls would take turns standing in front of the doors. This situation also required girls engaged in drug use to cooperatively block the doorways for friends, chatting amiably and fixing makeup and hair while a friend sniffed a quick line of cocaine or injected heroin in a spot covered by her costume. These bathrooms were also reputed to be equipped with cameras. Some long-term employees used non-spaces, like the ice machine closet or the mostly unused enclosed staircase, to temporarily disappear together to share cocaine and other more expensive drugs. Many women told me the tricks they used in hopes of avoiding the system of surveillance, as well as to avoid the social obligation to share their drugs more widely.

B-DRINKING AS SERVICE WORK IN A TOURISM ECONOMY

A thematic analysis of my informal interviews with Bourbon Street strippers and waitresses that began in the earliest phases of my research serendipitously directed my interest to the importance of B-drinking in the sexual economy of the French Quarter. Based on observations and interview research on French Quarter strip clubs, I suggested in 1993 that as much as 60 percent of the income of dancers on Bourbon Street had been earned through B-drinking (Demovic 1993, 23). Far from being only a creature of history, nightclub employees had continued to use this strategy to earn higher incomes than those received by other entry-level workers in the French Quarter. Years of conversations with tourism workers revealed that B-drinkers earned incomes comparable to some of the highest-earning waiters, or wait captains, in local fine dining—but without the requirement of years of experience. Indeed, several of my interview participants moved fluidly between the strip club and other kinds of work in the tourism economy, such as bartending and waitstaff, during the 1990s. Many more stay in the service industry as bartenders or waitstaff when they feel they are too old to continue as dancers, or when their incomes drop with age.

In most cases, women I interviewed engaged in such stigmatized work because it was an advancement strategy rather than a strategy of last resort. This is congruent with research elsewhere on sex workers who work for visiting populations like tourists or migrant workers. Denise Brennan's study in the Dominican Republic (2004) describes independent sex workers developing complex relationships with international tourists with a hope of emigrating or earning enough capital to start a business. In an ethnohistorical study of colonial Nairobi, Kenya, Louise White (1990) uncovered a class of migrant prostitutes who strategized to accumulate capital and purchase homes by providing services to migrant, laboring men who desired culturally appropriate feminine erotic and emotional labor. Likewise, providing erotic conversation and entertainment for visiting tourists in New Orleans allowed women to earn high incomes by tapping into money pumped into the city's economy by tourists, conventioneers, and business travelers. Focusing on visitors allowed for higher incomes without the problems caused by expectation for long-term relationships that could happen with local customers. Compared to other jobs in the service industry, these clubs were perceived to promise higher potential earnings, and the work paid cash at the end of each shift. On most evenings, most of the women on Bourbon Street I interviewed saw their work as a good fit.

A CITY FACING DISJUNCTURE

When we think about a map of American holiday making, or search a list of top-tiered convention destinations, New Orleans is a place of importance in terms of flows of travelers, attracting more than 9 million visitors per year. Louisiana is famous as a unique destination, yet despite its central positioning in the national (and even international) idea of the American touristscape it still occupies a weak position within the American financescape. In terms of flows of currencies and capital in America, New Orleans lies at the margins. New Orleans' host/guest relations are an example of the sorts of disjuncture between flows that Arjun Appadurai discusses in his broader work on how we experience places like New Orleans in the age of globalization (1990).

It is a poor city that is a notorious tourism destination; it is a landscape where America's very poor directly serves its very rich. Tourism in New Orleans, like other Caribbean destinations, is fraught with social and economic inequalities between its guests and those who work providing services. As an illustrative example, in one high end hotel in 1993 I observed employees (whether office workers or maids) submitting to having their handbags searched for hotel property each day as they departed work. Discarded bouquets of loose orchids and roses, changed daily in the hotel's ballrooms and hallways elaborate floral displays, could festoon the desk of a secretary typing convention contracts for minimum wage. Yet even discarded bouquets were not allowed to leave the premises past the guard at the employee entrance.

Despite New Orleans' role in American holiday making, it may be equally well known in the post-Katrina mediascape for the poverty of its residents. Between 2002 and 2004 Louisiana had the fourth highest poverty rate in the nation, according to U.S. Census Bureau Current Population Reports 2005. According to the 2015 American Community Survey, the people of the city of New Orleans have even lower median household incomes than found across the broader state. While 19.6 percent of the inhabitants of Louisiana lived in poverty between 2011 and 2015, during the same time within New Orleans 27 percent of persons were reported to be living in poverty (United States Census Bureau, QuickFacts, New Orleans). Louisiana is among the poorest of states in the union, and tourism accounts for 14.5 percent of jobs, more than any other industry (Williams et al. 2013). Speaking to the significance of the industry for the city, the Louisiana Department of Culture, Recreation & Tourism stated that "tourism is the economic engine of New Orleans, accounting for 35 percent of the City of New Orleans' annual operating budget" (2006–7 Marketing Plan).

According to geographer Campanella's (2008) timeline of the city, as early as 1922 the Association of Commerce Convention and Tourism Bureau attempted to counter the city's reputation as indolent and anti-business by promoting the slogan "America's Most Interesting City" as a replacement for the more passive sounding "City That Care Forgot" (2008, 46). The current boom in convention tourism began in the late 1960s, continued to grow in the 1980s, and further took off after 1995. In 1968 the convention trade was nurtured by the construction of the Rivergate Exhibition Hall at the intersection of Canal and Poydras (Campanella 2008, 55) [later demolished for the construction of the current Harrah's casino which opened in 1999 (62)]. The 100th anniversary of the World's Industrial and Cotton Centennial Exposition at Audubon Park was celebrated in 1984 along with the Louisiana World Exposition, held along present-day Convention Center Boulevard. At that time, the city's Warehouse District was revitalized into more hotels, condominiums, and art galleries. The exposition structures were later converted into the Ernest N. Morial Convention Center (59).

The 1980s also saw increased development of the riverfront, including the Jax Brewery building, in areas previously dominated by maritime businesses. The current Riverfront Streetcar, built in 1988, was the first new line built in decades, meant to make movement between the new convention center and the French Quarter easier (Campanella 2008, 59). At the same time, many white-collar jobs associated with petroleum companies moved to Houston while blue-collar jobs declined with increased mechanization of the city's famous ports. Campanella cites 1995 as the year that New Orleans began its tourism "renaissance," completing its transformation into a city dependent on the "tourism-oriented service economy" (2008, 61). This renaissance overlaps with my period of participant observation in the tourism economy.

The French Quarter is the oldest neighborhood in New Orleans and is today a National Historic Landmark. It is a vibrant and constantly performing heritage tourism site that draws swarms of independent tourists and hosts a healthy convention-scale tourism industry. Tourists go to eat, drink, and hear live music in the nightclubs that pepper the neighborhood alongside the strip clubs. The pieced-together remains of the contemporary Vieux Carré, or Old Quarter, are characterized by buildings that reflect the architecture of the Spanish colonial era. Fewer examples of the earlier French architecture remain, as most were destroyed during the city's historic fires of 1788 and 1794. Like the preservation of architecture that characterizes the city, many businesses maintain reputations that at least in part rely on notions of what is historic or authentic to the place. Somewhat parallel to the B-girl, the salesmen of the iconic hot-dog carts selling Lucky Dogs have been cemented into the image of the neighborhood even further by their role in the Pulitzer-winning novel *A Confederacy of Dunces* by John Kennedy Toole, published posthumously in 1980. Toole's characters are modern additions to Howard Jacobs' already growing list of "Droll Souls" (1973) who contribute to the performance of the place.

During the mid-1990s many entertainers I knew engaged in (re)productions of Bourbon Street professional identities that aimed for a pastiche of the glitz of 1940s and 1950s stripper styles, emulating a past when B-drinking striptease dancers were at their heyday of using elaborate props and constructing unique stage acts. Colorful feather boas and sequined gowns appeared in juxtaposition to black leather, vinyl, and the gothic fashions of the 1990s. Even today, dancers and B-girls whom I follow from the 1990s era send me social media posts of vintage glamour shots from *dame noire* of the 1920s and 1940s. "I wish I could dance in a giant champagne glass," one of my participants said, while we flipped through a set of vintage burlesque photos.

B-GIRLS FINDING THEIR PLACE IN CONVENTION TOURISM

A stranger to Bourbon Street dressing rooms might be surprised by the specialized knowledge that service workers in strip clubs have of the behaviors and schedules of professionals attending conferences in the city. Philippe Bourgois' (1996) ethnography of marginalized employment in New York (on crack-cocaine dealers in East Harlem) emphasizes that a lack of access to social capital limits his informants' abilities to excel in the formal, legal economy. Bourgois reports that those successful in the shadow economy often lack certain kinds of cultural capital necessary for success in the formal economy. His informants find it difficult to navigate the paperwork of bureaucracy and licensing for a formal business and inevitably fail to succeed in formal economic employment.

Similarly, some women who pursue work on Bourbon Street come from social backgrounds that do not provide them with the cultural capital needed to navigate careers in business or technology. However, as a group they successfully strategize to plan their schedules according to (and reasonably predict) the spending habits of groups of professionals arriving for conferences. It is their profession to observe and discover variations in behavior between the different categories of regular visitors, which they can exploit to their own advantage and maximize their income for their efforts. B-drinkers were often experts in tracking the activities of the formal tourist economy and paid careful attention to things like published convention calendars.

> The best money? Conventioneers. For many years I would get the convention schedule, add up how many people are gonna be there, OK cross out the Baptists, cross out the school conventions, and then total up for every day how many people are going to be there. And I have that on my calendar. And that's how I decide whether I am going to work or not.
>
> And the other thing you can do is they have those Bourbon Street webcams. So if you're like "oh there's only, you know, 3,000 people in town tonight, and its summer," well you can look on the Bourbon Street cams, and see how they look. Are they out tonight? Because sometimes the convention will have an activity and they're not on the street. Or, you can look how well-heeled they are. You can get a feel for the crowd. Jennifer, 2014 interview

Although in the early 1990s the Bourbon Street webcams were not yet a reality, the New Orleans Convention and Visitors Bureau Convention Calendar in paper form was kept behind the bar and shared with me by Max, a bartender on the busiest stretch of Bourbon Street. Max had a complex and evolving personal system of evaluating the upcoming month of work. He knew how crowded the French Quarter was expected to be and drew on his experience of the spending habits of those types of conferences. Associations of criminal defense lawyers, he told me, were likely to spend money on bottles late into the night. In contrast, an association of academics in the liberal arts would usually buy little more than the one drink for themselves and required for admission, or more often only peek through the doorway. Possibly, he said, a few solo men would come in late in the evening. Bartenders like Max watch the schedule from a different perspective than dancers.

Max shared his excitement when the conference goers were known for tipping, because waitresses who sold drinks to customers (for themselves to consume and for the company of B-drinkers) were expected to tip out to him. Waitresses in this bar earned no hourly pay; they earned only 85 percent of their tips, 10 percent of their total was owed to the bartender and 5 percent to the doorman/bouncer. Bartenders fared a bit better, earning $50 a shift plus tips at the bar and the expected tip out of 10 percent from each waitress.

Max carefully estimated sales throughout the evening as he poured drinks and collected money from each waitress for the cash register. He watched interactions between the waitstaff and customers and paid close attention to complaints about customers that had not tipped and arguments between waitresses and B-drinkers who had not pushed their customers to tip. He paid close attention when girls traded singles for larger bills from the cash registers and when customers changed large bills for smaller ones. At the end of a shift when the dancers were paid, they would estimate the percentage they owed to the doorman based on a suggestion of between 5 and 10 percent, and if they had needed the bartender to monitor the time for them while they were "on a bottle" they would give him a few extra dollars—so Max was also invested in monitoring the activities in the VIP room beyond calculating his cut of the waitresses' tips. Everyone carefully and discreetly watched the flows of money.

When particularly high-spending groups (such as a convention of representatives of a large distillery) visited who tipped dancers very well, dancers whom a bartender had befriended would also more generously tip out to him. This was not a required part of the pay scheme in this club, but was more in the spirit of the broader notion of tipping in the United States—for prompt and attentive service. B-girls were responsible for seeing that waitresses were tipped by the customers, and waitresses were expected to tip the bartender his portion as a part of the general pay scheme. Success depends in part on waitresses soliciting for B-girls, bartenders providing drinks as they are ordered, and making change for customers (for stage tipping) from the cash register.

Important to the bartender, who is often also in charge of the dance schedule, was the size of the conference and their projected propensity to create foot traffic. Even for groups that were projected to spend very little, there needed to be enough dancers in the club at all times to maintain the appearance of a party in the making. Scheduling for large events of mainly pink-collar professionals, like nurses, would present the bartender with the challenge of enticing dancers to come in when they were assured to make very little income. Sometimes scheduling was more complicated, and many times early evening crowds were very poor projectors of the overall business of the night, presenting the challenge of convincing the dancers to stay until the end of the shift. Male-dominated groups, like Max's example of lawyers, might be on the streets by 9 pm. In contrast, more mixed-gender conferences would delay the heaviest business of men visiting the nightclub. Mixed-gendered business dinners in restaurants or convention events might occupy these groups early, and men as individuals or groups often attend the strip clubs as an after-hours event.

Unlike dancers who make up the majority of B-drinkers, bartenders generally work on a schedule as employees of the club. They do not have the freedom the performers do to skip work on days when poor attendance is expected. They often are the ones responsible for creating a "dance list" as performers and waitstaff arrive at the start of any shift. The dance list includes all the girls by stage name, in the order they are performing. The creation of the list requires balancing performances for variance; when possible, two bleached blondes, two athletic pole dancers, or two non-white women in sequence are all avoided so that a customer whose preferences do not match the first dancer he sees does not have confirmation bias that the club is a bad match for him on the appearance of the next entertainer.

The bartender or manager keeping the dance list must also be adept at predicting the size of the crowd. He is the one responsible for deciding when the club has enough dancers to provide variation in the show, but not so many that there will be conflicts over limited customers and access to money in the house. He is sometimes responsible to call workers and persuade them to come in if there are not enough women; this is an aspect of emotional labor that is required of successful management. I have seen bartenders and club owners in the early morning as customers in bars frequented by dancers charismatically trying to cajole women from other clubs to move to their own. This was a kind of work that owners preferred to require managers or bartenders to do, as most business owners carefully separated themselves both socially and legally from the act of soliciting young women into employment as dancer/B-drinkers. One bar owner explicitly told me of his disgust for the kinds of places the women he hired frequented in after-work groups after 4 am, which were admittedly dark, dirty, low-end bars. Bar owners themselves are generally of a different socioeconomic class than their employees, and so part of the bartender/manager's job is to serve as a mediator between the club and the contracted performers. Thus, management must also engage in emotional labor with contract workers, as a bartender/manager who is beloved by the women is able to persuade them to work toward the bar's interests. He is also more likely to remain on the job, as his income relies on tips from these same women.

In the mid-1990s some women who worked in the bars knew B-drinking was illegal, and when customers asked about the practice they had usually learned from other workers to deny that they were being paid. Illegal or not, a dancer or bartender loses money by admitting to being paid a commission on gifted drinks. A savvy customer who is told his companion is already receiving a commission from the bar is less likely to additionally tip her for her time, or at least will adjust his tip. The successful B-drinker would allow customers the illusion that she was interested in their conversation rather than in the income from their drinks. One way this was done was by insisting on only accepting "single," or "small" sized cocktails from the few repeat, local customers, signaling to them that they were different from tourists and that the relationship was more genuine. This brought those customers back, ensuring a couple of extra dollars each week for the dancer he most preferred. Such local customers could sometimes be found chatting with the bartender, with a small champagne waiting, while a B-girl retreated to the VIP room temporarily with a higher-paying customer. Local customers rarely bring any large windfall incomes to women in VIP transactions.

It is an integral part of this emotional labor that the customer be allowed the perception that he is simply buying a girl a drink as he would in any other bar. He is persuaded to hope (or at least suspend disbelief, as at a movie) that this time is different, and, after all, sometimes the girl must fall for someone. As I discussed in chapter 1, the use of the gift of a drink to initiate a social contract of potential intimacy between strangers in a nightclub is common gendered behavior in the United States and Europe. The sexual script of reluctance on the part of the female, who loses her inhibitions and slowly gives in to her desires with mild intoxication, is performed over and over every night by B-drinkers. It is the reality of gendered interactions in American bars that stimulates a market for the striptease dancer's intimate performance as a B-girl after she exits the stage.

But the B-girl rarely actually loses her inhibitions, and her erotic performance usually ends with the purchase of the last drink. She may disappear into the dressing room until the customer leaves, reappear in a different wig and become invisible to her puzzled customer, or even leave out the backdoor if necessary. The barroom ritual of alcohol consumption and seduction is inverted. It does not empower the man, with an intoxicated woman persuaded toward further intimacy. Instead it reveals the power the woman has had over the carefully scripted performance, as the customer waiting for the B-girl to return from powdering her nose realizes she has given him the slip.

This is the scripted performance of the B-drinker. It has been part of being a good sport in an authentic run on Bourbon Street, for a soldier, for a tourist, for a convention goer, or for a real, authentic New Orleans' boy like Howard Jacobs. The same suspension of disbelief necessary for the enjoyment of any theatrical performance, or to receive services from those whose job is to convince us that they care (nurses, stewards, therapists), is a critical part of the B-drinking encounters. The hustle is precisely what is for sale.

THE HUSTLE

At the turn of the twenty-first century, my interviews revealed that many striptease dancers in the French Quarter agreed to B-drink because it was a requirement for employment within the clubs. Waitresses and bartenders also engaged in B-drinking, either regularly or occasionally when the opportunity to do so was present. The lucrative income generated from increased liquor sales motivated bar owners to require this work. The B-girls engage in a form of emotional labor, sitting with customers and striking up small talk, political talk, sexual talk, really any sort of talk that the customer finds entertaining enough to want to keep her company. He is then asked, rarely by the girl herself, if he would like to buy her a drink. This was as consistent in 1995 as in the past, as reflected by the strategies revealed in appeals to the law discussed in chapter 3.

One customer, searching for a girl to take to the VIP room, described the stage performances to me as "a sort of advertisement for the available girls." The stage performance allows the customer to see the different women, interact with them, and indicate through tipping or eye contact that he would like her to sit with him. Once she exits the stage, he finds that to keep her company he must pay for her drinks. The higher the price of the drink he purchases, the longer the period will be and the more secluded their conversation and potential private show. For the big-spender ("sluffing off his spondulics" as Jacobs said), a quiet booth with a beaded curtain may (or may not) lead to a lap dance (in which the dancer grinds her hips, pelvis, or buttocks against the seated customer). Waitresses complained about cleaning up bar napkins in the VIP room when men had received hand jobs in exchange for largess in the purchase of the B-drinker's time through buying champagne bottles. Such customers would also be expected to tip heavily—to the girl, to the waitress, to the bartender, all in unfixed amounts to keep the secrecy of the space he has purchased in the "Champagne Room," the "VIP Room," or the "Gentlemen's Lounge."

Some B-drinkers used the champagne room simply for table dancing, and some were known by the other girls to be willing to have more sexual contact. In my interviews, I found that most had been willing to seek out more cooperative B-drinkers for any customer who expected some additional services outside of her own boundaries, whether sexual or otherwise. There is a surprisingly broad range of activities that are requested in the VIP room. One B-girl I knew regularly had her toenails painted by customers who had purchased her drinks.

Legally defined acts of prostitution are not a part of every B-girl's repertoire, though not uncommon. I have talked to B-girls who provide sexual services, perform acts of domination, engage as temporary girlfriends, or even allow clients to cut their hair in exchange for the income they earn on drinks. What is common to all is soliciting drinks for commission, a job that during the 1990s consistently and easily brought each girl a modal income of roughly $100 a night, with lows of $40 and highs around $300. The highest nightly income reported to me was $700.

During this time, it was clear to me that almost every dancer in my sample earned a higher percentage of her income from B-drinking than from striptease dance. The formal, legal aspect of her work, erotic dance, was often overshadowed by the illegal work of soliciting drinks. Women who were in the VIP room could be skipped when their turn came up on the dance list, as their incomes and the club's income would go down if they left customers buying bottles to perform on the stage. Despite this, very few of my informants had ever observed, feared, or in fact even knew of anyone who had been arrested for this crime. It is perhaps this infrequency of arrest that has led researchers to believe there were a few B-drinkers in the French Quarter among a sea of non-B-girls. But at the turn of the twenty-first century this was not the case.

Women who engaged in B-drinking sometimes moved in and out of various forms of engagement in sex work in the French Quarter. Rather than agency or sex work being simple on/off switches, these workers conveyed a message of complex decision making in which individual acts of sex work were negotiated with considerations of risk, potential profit, seasonality, and often with a deep understanding of gender stratification in the United States. Rather than a simple matter of choice or cultural coercion, the decision of whether and how to engage in sex work was viewed by the workers themselves as a complex process. Through the voices of these workers, the messy issue of incompatible feminist theoretical models disappears and is replaced by pragmatic decision making in a gendered world.

B-GIRL CONSTRUCTIONS OF WORK

> Well, as far as the job, there is one part, as the stage show. And that's where you really are an entertainer. Unfortunately that doesn't—that pays such a small part of the overall thing—that I've become very lazy on the stage.—Lana, interview, 2014

> I've never made extra money off of doing pole tricks. I actually made more money off of just going up to people and holding out my garter and asking for a dollar than doing really fabulous performances on stage. And then, to me the biggest part is, you know, being a hustler. Talking to people.—Dixie, interview, 2014

All three topless clubs in my sample in urban New Orleans paid women to B-drink, although it was a less important strategy in some clubs than others.

Regardless of the economic importance of B-drinking, few informants use the term "B-drinker" to describe themselves. Instead, this is a legal term that has structured the way B-girls are treated and imagined.

In a structured survey (table 4.1), ten women who had worked in the clubs for three or more years answered a question asking them to describe "what sort of work or association did/do you have with employment on Bourbon Street." Participants were not limited in the number of ways they could self-identify, and many of them identified in multiple categories.

In response to a reviewer who disagreed with the notion that B-drinking was a form of sex work, I returned to a group of my informants. Although only one had used the term "sex work" herself, all four strongly agreed that the services they had sold as B-drinkers were erotic in nature. Additionally, they pointed out that some women from the 1990s until the present used VIP rooms for transactions in which cash or bottle purchases were exchanged for acts legally defined in Louisiana as either prostitution or crimes against nature; one reason a neutral term such as sex work is necessary to discuss sexual commerce is because state laws that define prostitution vary. Al-

Table 4.1. Responses of Ten B-drinking Informants' Self-Descriptions of Work on Bourbon Street

Term given by participant	*# of Participants Generating the Term*	*TOTALS of each category of work*
Dancer	4	
Topless dancer	1	
Exotic dancer	1	
Stripper	2	
Entertainer	1	9 of 10 identified as *entertainers*
Dancer/b drinker	1	
B-drinker	1	
Shot girl	1	
Waitress	2	
Bartender	2	7 of 10 identified as *selling alcohol*
Bar/restaurant manager	1	1 identified as *managing sales*
Private independent parties	1	1 identified as *doing outside sex work*

Created by author

though many entertainers (such as movie actresses) *sometimes* engage in performances that are sexual in nature, the B-girls performance *is always* erotic in nature.

Participants in my later interviews revealed to me that many clubs not only in the French Quarter but across the greater metropolitan area hired female B-drinkers. Although my own observations were limited to a handful of clubs within the French Quarter, women told me that B-drinking skills were useful in clubs as far east as Florida. Some dancers would go to Florida during the summer months because of seasonal shifts in tourist arrivals. Archives suggest that B-drinking has also historically been a practice in Bossier City, Louisiana—although none of my informants ever mentioned clubs from that region. A few of the women I spoke with, particularly older dancers that I met in the 1990s, worked even further afield, in Las Vegas and Hawai'i. These women were able to transfer their B-drinking skills between these regions, which leads me to believe that similar systems were probably in place.

All the women I interviewed who worked on Bourbon Street were familiar with the practice of B-drinking, and all had participated in it. Interviews and observations suggested that women employed on Bourbon Street received between seven and twenty-two drinks per night while working as dancers. In one club in my sample, a male bartender was paid to B-drink. In the same bar, as already mentioned, at least one extremely successful B-drinker was transgender. I doubt this was an isolated occurrence, as there was a bar (not included in my observations) at the start of my research period that specialized in cross-dressing, and there is little reason to believe that those individuals did not integrate into other existing strip clubs when the place closed down.

Both on and off the main tourism section of Bourbon Street, dancers often received no hourly pay at all. Instead, they relied only on their tips and some percentage (commissions) of club income acquired through B-drinking and performing private dances. In two clubs, participants reported that a failure to solicit a quota of a minimum number of drinks, set and enforced by the management, resulted in various sanctions for those dancers. Sanctions at one club included being suspended from work or fired. Another club offered an hourly wage of five dollars, but only for women who met the minimum drink quota of eight drinks per shift. Those who did not meet the quota would receive their earned commissions and tips, but were not paid the conditional hourly wage.

Such policies are devastating during slow, non-tourist seasons, when five dollars an hour can represent more than half of the shift income of the employee. This contributes to intensification in competition for what dollars are spent on Bourbon Street during the hot summer months. More dancers were willing to engage in out-call prostitution during the slowest tourism months of June until September out of necessity to meet their financial obligations than at other times of the year. This is also the time of year when dancers are most likely to have fights in the dressing room, which can be brutal. In one instance, I narrowly avoided being toppled over by two fighting dancers when I entered a club dressing room. The perceived "theft" of a customer ended in one woman's body piercing being ripped out of her stomach while she held the other woman by her hair.

In the latter half of the 1990s, some of the most upscale bars on Bourbon Street began the practice of charging dancers a "stage fee" of between $50 and $150 for the opportunity to dance in the club for a night. These employees reported that they were still paid to drink, and in some instances their stage fee could be "waived" for the evening if they sold (and ostensibly drank or otherwise disposed of[3]) a quota of cocktails and bottles. Such stage charges did not mean additional services were being provided for dancers (such as the hairdressers, etc., mentioned by Judith Hanna [2012] in discussing such fees). Instead, these fees were imposed as a part of the constant struggle between B-drinkers who took cash directly from clients and nightclub staff and owners who depended on B-drinkers earning high incomes on VIP bottle transactions to transfer more of that income into the business' cash register.

The dancer/B-drinker is incentivized to hide her income, requesting large tips rather than additional drinks or bottles. Unlike purchased drinks, tipping is a silent payment directly from the customer to the performer, and she may or may not choose to tip out to other staff honestly. Thus, she can minimize her payouts to other staff, and staff are incentivized to keep careful track of her every move and transaction. In blatant examples of such struggles, bar managers sometimes attempted to seize portions of "tips" deemed to be an unfair distribution of incomes.

In one example, a nightclub owner confronted a performer who he had seen receiving a one-hundred-dollar tip and demanded she divide the tip, paying half into the cash register or lose her job. Another nightclub owner regularly complained to me about the electricity bill for the club, gesturing angrily at the elaborate colored lighting and disco ball turning above the stage and the wide-open door in the air-conditioned club to make his point about how such dancers mistreated him. He used these examples to complain about the costs he absorbed that were invisible to the entertainers, and he passionately defended his actions by saying that excessive tipping to the women cheated him of money that should have gone through the cash register in the form of the purchase of more, or larger, B-drinks.

Some participants interviewed in depth during my 2013–2014 visits reported to my surprise that during the 1990s they had been paid to B-drink in non-strip tease bars while working as bartenders in and around the French Quarter at other tourist-oriented businesses in the city. Most often, for my participants, B-drinking was paired with semi-nude dancing in the French Quarter. However, bartending in inexpensive bars as well as at least one upscale hotel bar were other formal jobs that included the potential for additional informal income through B-drinking. My focus on Bourbon Street clubs didn't include such bars, so the extent of the practice is beyond the scope of my study.

NOTES

1. Campanella (2014, 267–268) represents foot traffic in two ways in his book on Bourbon Street. One is a map that represents flows of movement of pedestrians who tend to emanate from Canal Street (the site of the streetcar stops for the Quarter, as well as many hotels) down either Bourbon, Royal, or Decatur Street toward the back of the St. Louis Cathedral and then moving parallel to the river toward the French Market. This map (267) provides data for a locally hosted Superbowl Sunday, which estimates traffic in front of all three clubs under study to be in the highest range based on his survey, of "60–80 pedestrians/minute." He also charts foot traffic along Bourbon Street by blocks (268). I here use his data on rates of foot traffic for the two blocks on which the Bourbon Street clubs under study were located and averaged them. I hope this gives a general sense of the high visibility of the clubs without further identifying them.

2. Only a handful of regulars visit The Den because there are less expensive clubs with less prestigious/expensive locations relative to the tourists' points of interest in the French Quarter. I maintain throughout this text that strategies that function in the sexual economy of tourism are very different from those that would work in a bar where regular customers want a more realistic performance of a relationship rather than a tourist experience.

3. Not all B-drinks are consumed. There are a wide variety of methods employed by B-drinkers to discard their drinks surreptitiously. A girl may carry a full glass to the bathroom, where she can dump it and replace it with water, for example. A common strategy was to simply dump part of the contents of the champagne bottle into the ice bucket while filling a glass or while distracting the customer. This was something women also did for each other, both other B-drinkers and waitresses.

Chapter Five

Understanding the Perspectives of B-girls at the Turn of the Century

BEGINNING WORK AS A B-DRINKER

Multiple discussions I had with women about how they began work in the nightclubs focused on the ease and speed with which they could gain employment. Financially unstable women who arrived in the city with very little or no money could begin work immediately. Some women already had experience dancing in other cities and moved to New Orleans with the idea of continuing their work as dancers. Performers could travel in and out of the cities of the southeast seasonally, with shifts in the seasonality of tourism. This mobility is consistent with burlesque dance since its beginnings, as the earliest burlesque performers tended to move from city to city to perform in a circuit of venues in different towns in rotation, what Leslie Zemeckis (2013, ch. 23 p2) refers to as "the wheel."

The hiring process for a new girl seemed to be as simple as a quick look at a driver's license to determine age eligibility (or at least a false identification card, reducing the manager's liability in employing underage girls or undocumented workers on his shift).

> I had planned to go work in restaurants; we got there (to New Orleans) really late at night like 2 am. I guess I was 24 at this point. And we got to the Quarter about noon the next day, and just started walking into places. Because back then, you didn't apply for anything online, you had to show up. So anything in the service industry, you wouldn't just drop a resume at a restaurant back then. You would fill out an application and talk to the manager. So I went out and just kinda started hitting the streets and I was like, you know we had a hotel, we had $150 in our pocket, and we needed to make money for the next night's hotel, and food and everything. . . . Jade, interview, 2013

Figure 5.1. Word Frequency Cloud Based on Interviews about "First Work on Bourbon Street" (created with NVivo 10). Created by author

Jade continued, "I had been in New Orleans not even 24 hours and I was going in to work. Which is really cool." For women seeking to relocate with very little money, the nightclub environment is a fast solution.[1] A woman could walk in at the start of a shift seeking employment and leave 9 hours later with whatever income she earned that evening in cash. This system also meant that new women, unknown to the established group of performers who remained year-round, were treated with suspicion and not often left alone in the dressing room to prevent theft. Women reported that they often traveled in small groups of two or three and would arrive in any new city for a season of work together.

Even classified as contract workers, many women worked for years for one club on Bourbon Street, making up a core group of trusted insiders. This is very different from Susan Dewey's ethnography of a striptease club she calls Vixens in the "rust belt" (2011, xii). Dewey states that most of the approximately fifty women she interviewed "left the job within just a few months—which is typical in an industry that benefits from the greater variety of performers resulting from high employee turnover." In my sample of 71 women who worked on Bourbon Street in the 1990s, most worked on the strip, and often in the same club, for more a year—some for over a decade. In the tourist environment, a variety of dancers on a given night is essential, but variety between nights does not benefit the industry. On Bourbon Street the customers change from night to night and so the performers do not have to. Further, a longer-term employee understands the cultural geography of the bar described in chapter 4 and can more efficiently navigate the club's strategies, spaces, and their meanings.

The women across my sample developed a trusted position within the community through their history of either staying or returning seasonally to work with one club on Bourbon Street. Returning to the same club made moves from one city to another easier for performers, as they arrived with an insider status to the Goffmanesque deep "back stage" spaces of the dressing room and were easily reintegrated through known rituals, such as shift drinks (discussed in chapter 1) consumed with coworkers who knew them (or sometimes only their reputation, communicated by older dancers). Contract performers also cultivated (or at least performed situationally) a sense of limited loyalty to certain managers or club owners, relationships which could be important when large sums of money were needed in a crisis.

Club owners and managers sometimes made short-term loans to contract performers who they had employed repeatedly and whose movements were predictable. Like other informal industry (or underground/shadow economy) workers, most B-girls do not report their incomes and therefore have no obvious means of repaying loans. Women employed as B-drinkers are often not in a financial position to receive credit outside of such personal loans, and management knows this. Although some have enough legal income from other sources to have established a minimal range of credit, generally the B-drinkers I knew paid most of their bills in cash. Most of the B-girls and dancers I interviewed did not have access to banks or credit cards and when money was short would use predatory lending institutions such as those that take car titles as collateral. They must either save or borrow from friends when they needed deposits for apartments, medical procedures, or other expensive bills. In one case, a young woman I knew received a large loan from a club manager for an emergency surgery for endometrial cancer that was probably lifesaving. Without medical insurance or even a formal income, a contract performer has many reasons to develop long-term ties with one bar or one manager.

From the perspective of a bar owner or manager, such loans are not an altruistic act. Large loans carried with them the expectation that the recipient would behave as an employee, rather than an independent contractor, until she had paid back the loan. She might be asked to work many scheduled days without breaks, change her style of costume or dance, or any other number of commitments usually not required of contract workers in exchange for the loan. In the case of a large loan, like the one needed for the surgery mentioned previously, this can result in an informal contract of indentured labor with most or all income withheld each night until the loan is repaid. Such arrangements between club owners and contract workers are not discussed openly.

> You know, I was always careful in how I worked with (bar owner). I really needed money one time, because I was pregnant and wanted an abortion. I was really upset, and I didn't know how I was going to get the money. So I went and talked to him, because I knew he had given girls money in the past. It was really awful, I was crying and spent all this time talking to him about it. Anyways he told me to come back the next day and he would loan me the three hundred dollars. But that night I found Jade in the Quarter, and she had had some really good nights and loaned me the money so that was a lot more comfortable for me. And I went back the next day and told him I didn't need the money and he was *fucking pissed*! He was so mad about it he wouldn't let me work that night. I don't know. It was weird. But I kind of felt like I had dodged a bullet somehow. Like, why would it make him so mad?—Sadie, interview, 2014

Judith Lynne Hanna, an expert on the anthropology of dance and the First Amendment protection of striptease, claims that women who have negative experiences in strip clubs are few, and that such women are sought out and used by Christian Rights Activists to testify publicly about harm within the workplace of the strip club. She contextualizes this as part of a broader effort by the Christian Right to close such clubs. In describing the working conditions of contemporary gentlemen's clubs, Hanna wonders "what income, loans, educational tuition and college expenses, daily fresh flowers, financial counseling, and training in work habits the CR (Christian Right) Activist organizations provide" (2012, 209), painting a positive picture of such benefits. I have personally received flowers in The Den, loans are indeed sometimes made to contract workers, and one dancer/B-drinker I knew had her community college tuition paid by a manager (as part of an ongoing sexual relationship). The situation between employers and workers is fraught—it is not a binary of all good or all bad when considering the benefits and problems of Bourbon Street working conditions.

I suspect Hanna's enthusiasm in defending striptease is born of her work as an expert witness. Her role as an expert on dance in the courtroom is to minimize *any* claims to harm, in part to keep strip clubs from being closed. Understandably, as an expert witness on dance, her own gaze is focused on the stage and other sites of dance. My perspective as an economic anthropologist directs my gaze elsewhere; it is the economic transactions happening in the bar and not the right to dance that is my primary focus. Although I agree with Hanna about the right to communicate through dance, in many ways striptease dance itself is only the backdrop of contemporary B-drinking, which is the focus of this book.

NETWORKS OF SUPPORT

Coming back to New Orleans in 1995 after a summer away, I am driving around the French Quarter looking at for-rent signs in windows and trying to find an address copied down from a newspaper ad on a smudged piece of paper. I've driven back to New Orleans from Chicago with a friend, and we are hoping to find an apartment together. We decide to park the car in an empty spot, willing to gamble on the possibility of having the windows broken and our stuff stolen because it's still daylight and foot traffic is heavy.

As we are walking around the Quarter my friend is looking a little overwhelmed. We meet a street performer, a tall man in a clown costume, and he strikes up a conversation with her. She is new to the city, and naïve, and it takes her only a few minutes to reveal our entire predicament of temporary homelessness to this stranger. She is excited when he makes an offer to let us stay at his place. While I try to gently disengage my friend, the clown walks her (us) a few buildings down the street to a doorway which he says is his apartment. "Just go inside, and go in the kitchen. There is a bunch of food in the fridge. If you just cook dinner, you can stay as long as you want to." She turns to me with a questioning look, and I know she is thinking about the two hundred dollars we have between us and the cost of a hotel and thinks this seems like a viable offer. "The door uses a key from the inside and out, and I only have one key, so I will have to lock you in. But I'll be back as soon as I am done working," he continues.

"Are you OUT of your mind?" I tell my friend and drag her away from our new acquaintance. He seems very disappointed and yells a few misogynistic remarks at me as we rush away. At that moment, I am suddenly happy to find Jimmy leaning up against a building with a plastic go-cup of beer in his hand. Jimmy has worked as a barker and doorman in one of the Bourbon Street nightclubs on and off, and I have known him for a couple of years.

Jimmy is short, stocky, and has a lot of jailhouse tattoos. I've heard him complain about being "stuck" in his job as a doorman, although he never tells any of the details of his story. The club owner where he works seems to know more and is alternately sympathetic and cruel to Jimmy. He sometimes forces the dancers to tip Jimmy the 5 percent he is technically owed, which adds between forty and one hundred dollars cash to the five dollars an hour he earns for a shift. Although he legally is not allowed to call out to people on the street, Jimmy is well known for his lewd comments meant to entice customers into the doorway. "No chicks with dicks, no weenies in the bikinis, no sluts with nuts! All nude, all girls, all you want," he calls out from doorways at least five nights a week. He is not very attractive, and even when they are bored the dancers are more likely to tease him than to flirt with him. Twice, when police have been called to the club, I've watched Jimmy calmly walk through the club and out the back-office door, escaping the club through the courtyard and walkway and avoiding presenting any identification. I suspect there was some situation that made a cash-paid job a necessity.

I tell Jimmy about the clown, and he laughs at my friend and warns her that she shouldn't let anyone lock her into any buildings—sage advice. He is hungry, and he asks us to come back to his place and eat dinner with him. We agree and start walking through the French Quarter and away from my car. I am a little worried about my belongings, but the ethnographer within me decides we need to at least visit for a few minutes. Soon, we emerge on Rampart Street, a street that marks the edge of the Quarter and the beginning of the neighborhood called Treme. I think about all the times leading walking tours that I had advised tourists not to walk on Rampart Street after dark, even if they wanted to go to an address on that street.

Jimmy lives in a multi-story building with steps leading up to a small porch and entryway. There are older men, disheveled and smelly, smoking cigarettes and drinking forty-ounce bottles of malt liquor on the steps who stare at us. Jimmy gruffly tells them that we are his friends and to leave us alone, but I am a bit nervous walking past the men up the stairs. I am glad when we get to Jimmy's single room, which contains a table and one chair, a bed, a TV on a discarded dance box, and a mini fridge. He explains that he shares a bathroom with the rest of the guys on the floor but suggests that the bar down the street would be better for us if we need to use a ladies' room.

My friend and I sit cautiously on the edge of the bed, and Jimmy makes us cold sandwiches of Bunny brand white bread and cold, precooked chicken-fried-steak patties. We eat the sandwiches while chatting about how slow business has been and how hot the weather has been all summer long. "You don't miss anything leaving, Angel," he tells me. He offers to let us stay there for the night, and I gesture to the single full-sized bed and joke with him to "quit clowning with us." He laughs, and after another half hour of chitchat we decide to get back to the car and move out to a cheap hotel in the suburbs for the night. I tell Jimmy to let the club owner where he works know I am back in town and want to talk. He agrees, and we leave.

This moment has always stuck with me. I have never been poor enough to live in a flophouse like Jimmy. Jimmy also knows, as the doorman of a club, that he is the lowest-paid member of the nightclub's staff. Jimmy knows I work at a university, and he sometimes took part in teasing me about it. Yet the equalizing factor of belonging to this stigmatized community, the rituals created to move as a group between one's everyday identity and the identity of performance that happen every night on the floor of the club, the shared knowledge of lucrative and sometimes illegal activities all lead to a sense of strong community, even communitas (Turner, 1969) that develops among long-term coworkers. That sense of community was revealed in the sharing of those cold sandwiches from an almost empty fridge. Marginalized workers in the nightclubs of Bourbon Street form tight relationships that are important to both their economic success and their self-identities in a larger community that stigmatizes them.

Having a relationship with one club creates a bit of a financial safety net, even though the women are technically self-employed. That safety net comes in part from a relationship with management, but more importantly from the community of other employees and B-drinkers. It is sometimes the case that groups of women from one club share apartments or motel rooms, and women who migrate out of the city in the summer (when New Orleans suffers a tourism low but nearby Florida resort towns are busy) can thus more easily find temporary housing when they settle back in for the busier convention seasons.

In 2014, women who were then working for one of the large, corporate-owned strip-club chains talked to me about assisting one another in moving between two large southern cites as needed to staff the clubs during busy weeks, such as during professional football games. These moves to meet the needs of shifting tourism flows, because of conventions or seasonal events, are in part possible because of women who cooperate in providing shared temporary housing for visiting dancers from such sister clubs.

Because women are hired on the spot and are paid in cash at the end of each night, any circumstance that might require a fast, high income can lead women to enter the nightclub economy of Bourbon Street. Regardless of the initial event that leads to such employment, in every case the participants in my interviews indicated that they liked their work in the clubs.

> Most of the women I knew there were strong, independent, intelligent women—some of whom did it because they loved it. Others did it for some extra money, others did it to pay off their college bills, be able to afford to have their kids. Others did it because of drugs, and I think there were some girls who did it for all of the above reasons. —Sadie, interview, 2014

The themes of quick money and enjoying the work arise yet again as Tonya explains how she ended up returning to the clubs after a short break from work on the strip. She had left nightclub work to change her work environment, in part to reduce her consumption of alcohol. But for families that just survive on working-class incomes, any immediate and urgent need for cash can lead to a necessary return to the higher-income, higher-risk work in nightclubs.

> I ended up going to [named club] when I went back [returned to dancing]. My ex-husband had gotten put in jail and I needed money to get him out. So that was like, ok, well let me go work. And I worked two nights while he was in jail, and did so well, and really liked it that I just never left. And I ended up staying in the clubs for like 2 years after that at least, maybe three. —Interview, Tonya, 2014

Shifting between broad ranges of occupations across the service industry of the French Quarter is not uncommon. Women who are tired of working in the clubs often leave for other types of work. Camellia, for example, worked on and off as a fine-dining waitress. Waitressing in the French Quarter, however, is by no means unskilled labor. Fine-dining restaurant staff is highly hierarchized, with head waiters overseeing the work and gathering a higher percentage of tips than less-experienced waitstaff. Although in many parts of the country waitstaff jobs are less competitive, at four- and five-star restaurants in the tourism economy of the French Quarter such jobs pay well and are difficult to acquire. Unlike waitressing in a cocktail bar, applying to wait tables in fine dining requires a formal resume reflecting increasing experience.

For those who do work in restaurants, there are strict schedules of work times. Restaurant customers with reservations, sometimes made far in advance as part of a touristic agenda, expect timely and excellent service from those who wait upon them. This is far different from the strip clubs that employed B-girls, which did not require a set schedule and would sometimes allow a dancer to start work even hours late, depending upon business and the number of women in that night's lineup of performers. A few women who were viewed as particularly high earners could show up two hours late for the night shift on a regular basis, avoiding the low-income early evening hours and depending on their relationship with management to still be allowed to work later in the night. Management might complain about such women but were hesitant to refuse to let them work unless the house was already full, the dance lineup was already too long, or they could afford to send them home to make a point about power relations.

In clubs that don't have a day shift, at around 7 pm staff begin arriving on Bourbon Street. It is part of the manager's job to rush dancers and waitresses to get dressed and on the floor, but pressing too hard simply leads to dancers rebelling by appearing on stage for their first set in an awkward combination of stage makeup and street clothes; this and similar everyday acts of resistance allow contract employees to control the club in many ways from the dressing room.

There can also be problems if not enough dancers come to work, especially on a busy convention night. If not enough girls have arrived on time, a club might agree to hire an unknown dancer to begin at 9 pm for a shift that began at 7 pm, a cause of potential conflict among the regular group of dancers from the club. Although persistent lateness is problematic across various kinds of service-industry work, lateness is measured differently in fine-dining restaurants than in nightclubs. In restaurant work, fifteen minutes late may be grounds for potential dismissal. For women working as dancers and B-drinkers, lateness is measured in hours rather than minutes. Here, Amber talks about her difficulties in maintaining a waitressing job.

> I stayed in fine dining for a couple years. And I ended up going back to strip clubs because by that point I already had an issue with drugs. So . . . it was a way to make money, and also, it had gotten to the point where I was being late for my jobs a lot. And it was like I was always in trouble for being late. And rather than work somewhere that I had to be here this certain time, which I couldn't do, you know I went to the strip club to work, where I could be there when I wanted to work, where I could make a lot of money, and I had friends who worked there.—Amber, interview, 2013

Shifting roles and positions seen throughout the service industry, and between types of workplaces, also occurs *within* the nightclub. The Bureau of Labor Statistics suggests that in 2015 turnover rate within the hospitality industry was 72.1 percent, in comparison to an average rate of 45.9 percent in all private-sector work. These rates are still lower than turnover rates before the recession, with turnover in hospitality work at 80.7 percent in 2007 (National Restaurant Association 2016). In an article on employee turnover rates rising, the National Restaurant Association suggested that increases in the quit-rate of workers in the hospitality sector "indicat(ed) that workers are increasingly confident in the labor market and are willing to move to another job." This applies also to my own findings about women working in a continuously operating, tourism-driven entertainment district like Bourbon Street. It is quite possible on Bourbon Street to walk out of one job and into another on the same night.

It is not uncommon to see a woman initially hired as a club waitress change to dancing and B-drinking. Keeping a job as a waitress is more difficult than continuing as a dancer for many reasons. While dancers can take relatively long breaks, a waitress must spend even the slowest times posted at the door waiting to serve as hostess.

> At all the other customer service jobs I've had, I've been a grouch due to compulsory politeness, which I can hardly pull off under ideal conditions. Once I became a dancer, I had a choice. Waitresses have to kiss the ass of every person they wait on, even if the customer is a rich, demanding prick. If any customers at my club is rude or overly demanding, I can either refuse to speak to him or have him thrown out. This is a powerful position for women who have primarily been under the charge of men for most of their lives.

"Janet." *Rocket Queen* (independently published "zine," catalogued at the Newcomb Archives for Research on Women)

None of the three clubs observed in the 1990s hired separate hostesses, a practice more common today in larger clubs. The waitresses could serve in similar roles to barkers, working to entice customers to a chair through a performance of bawdy wit and hospitality once they were inside or near the doorway. Many people walk in and out of clubs, and a slow or dull waitress will lead to the loss of potential but hesitant customers. In many ways, the doorman or waitress of small strip clubs brought forward in time the tradition of burlesque performances including comic acts (see Zemeckis 2013 for an interview-based history of the transition from vaudeville to burlesque and the role of comic acts in each). I saw many instances of comic performances like Jimmy's lewd rhymes and less-than-subtle innuendo. One athletic young doorman used a broom as an extension of his arm to do cartwheels while miming sweeping the sidewalk and calling out to potential customers. Dancers also used lascivious humor in their acts, like Nile, who would lick matches, stick them to her nipples, and light them on stage when a customer came in with friends celebrating a birthday.

A Bourbon Street strip club could run with a skeleton staff of a doorman, a bartender who was willing to also manage the club, and a waitress. Once those employees were in place, it was never certain how many dancers would decide to come to work; although waitresses were also considered independent contractors by clubs (and received only tips), like bartenders they had to adhere to a work schedule or lose their positions. Thus, for the same reason employees leave fine dining for the clubs, they often leave waitressing work within the strip club for the relatively easier schedule of the dancer. Women would begin as waitstaff and then transition to dancers, sometimes by choice and sometimes because they had been fired as waitresses. During one observation, when the bartender and waitress had arrived before any dancers and a customer came in, the waitress was asked to dance a set (to the amusement of the customer, amused by her "one-gal show").

Bartender and waitress positions are scheduled, and presumably in some cases bartenders or managers were hired as actual employees of clubs who received tax documents rather than working as independent contractors. Although bartenders and doormen were men in most cases, in every case observed the waitstaff was entirely feminine (either women or transgender). Based on notes comparing incomes of 16 women in 1993, I concluded that the average income for waitresses was higher than that for dancers, although dancers often were not aware of this.

Management is eager to fill an open waitressing position because B-drinking relies on an intermediary to complete the sale. In some cases, if the customer sits at the bar, it can be the bartender who serves the drink. Some bartenders would ignore B-girls with empty glasses who did not ferry their customers to a table after one drink, as soliciting for B-drinks was viewed as the specialty of the waitress. By serving a customer and B-girl at the bar, the bartender may earn a few extra tips—but by doing so, he risks alienating the waitresses who tip him at the end of the night based on their overall earnings. In every club, most seats were at tables or around the stage. A waitress goes through a sort of probationary period to see if she is able to keep up the pace, particularly on busy convention nights. Tracy transitioned from waitressing to dancing because "you don't have to run around, meeting everyone at the door. You can just do your thing." Vicky expressed a similar sentiment about the ease of B-drinking and dancing relative to waitressing.

> I had been working in another club [waitressing in club without dancers], right down the street and I made really good money, but you know I was partying all the time and everything. My work was starting to restrict certain things . . . so we weren't making as much money. It was not as easy to go in and walk out with $200 like it was before. So, I started looking for other things and someone suggested 'Well why don't you come cocktail [waitress] at the [named strip club].' And I was like, well if I'm gonna work there I'm not gonna cocktail, I'm just gonna dance. You know? If I'm gonna work there I'm just gonna go for it. —Vicky, interview, 2014

DANCERS AND B-DRINKING

B-drinking is done by waitresses, bartenders, and most rarely by women who have come to work with the sole intention of B-drinking. Most B-drinking observed in the 1990s was done by topless dancers. Although women sometimes invested considerably in both their athletic dance performance and their unique costumes, my focus on B-drinking emerged as I learned that dance tips were not the source of highest income for women in these three clubs. Indeed, some high-earning B-drinkers invested very little effort into their stage performances.

In discussing their past work in French Quarter strip clubs, interview participants in 2012–2014 tended to first self-identify as dancers. For many of them, this was the favorite aspect of the work. "I don't think the dancing part was degrading, I actually enjoyed that," explained Sadie during an interview. For most of the women I interviewed, the dance component of the work was the focus of their occupational identities. This is interesting because the greater majority of their income came from B-drinking; tip money earned during dance sets made up a much smaller amount of their take-home pay.

Many of the women I interviewed who worked as semi-nude dancers reported a deep interest in the arts broadly and did view their stage time as an outlet for creativity and self-expression. In clubs where women could choose their own music, the mood of the dancer played an important role in performance through choice of costume and music selected.

> The actual performance, which for me you know, as an artist who was yet unfulfilled in my own artistry [laughing], you know at that time that was my great form of self-expression, and I had fun with the shows. I had a little original, I had my own little racket, not racket but . . . um . . . gimmick. . . . You know and I took it seriously . . . somebody wrote an article one time about the different clubs . . . and the ladies of [named club], and I remember the quote said "And the ladies over here dance like their *lives depend on it*" (laughing).—Mary, interview, 2014

> The whole stage thing . . . to me, working the stage is completely different than actually working with the clients, the customers. I really liked the stage, because I like to perform, I like music. —Jennifer, interview, 2013

Another potential source of income through dance in the 1990s and today is the sale of private dances. New Orleans law (Code 1956, § 5-76.2 Sec. 10-435) requires that "all shows . . . wherein a striptease or burlesque performance is conducted or engaged in, shall be performed entirely upon a stage or platform which shall be raised at least 18 inches above the floor."

In two of the three clubs observed, dancers were permitted by management to perform legal "table dances" that satisfied the 18-inch rule. In practice, individual girls chose to dance on the box, with one foot on the box, in the customers lap, or any number of negotiable positioning in between. This depended on the individual dancer and upon how lax management was in enforcing the rule. Though contact was illegal, women wearing any number of themed costumes, from black latex to school uniforms, some in only a "t-back" (a thick thong made of opaque material), performed contact erotic dance during private dances in the back VIP/champagne room of most clubs observed in the 1990s. Over a decade of interactions, I heard stories about owners creating and retracting house rules about when and if private dances could be sold, balancing the issues of liability if a girl were to fall off the box and be injured with the potential of being arrested if girls did not follow the law. Much of the time, private dances were only permitted if the customer had already purchased enough B-drinks (or a bottle of champagne) and could thus enter the VIP room. Once in the VIP room, the price of a private dance was negotiated with the performer.

There is a great deal of variation in how much contact was allowed by each dancer. Again, like working hours, individual women had a lot of agency in deciding where they drew their lines of what they did and did not permit in the context of a stage dance, a private dance, or time spent in the champagne room. Negotiations among women working a shift together about expected prices and acceptable levels of contact were sometimes a source of conflict.

> This was at [specific bar] and one guy [part of a group], like grabbed my boob, and I was up on stage. So I slapped him really hard, after throwing my garnet ring, which left a print in his forehead. And as soon as that happened he was pissed off. But, not all the girls were that way. Girls that I loved a great deal were mad at me for messing with their money. Because they knew that they could make so much money off these guys, and they were just about to go off to the back room, and they just didn't understand why I would react in such a violent way.—Scarlette, interview, 2013

Scarlette was an example of a dancer who strictly adhered to the 18-inch rule, not because it was the law (she knew contact was illegal, but none of the dancers interviewed were aware of the mandatory distance), but because of her personal limitations on physical contact with customers. She was also often one of the lowest-paid girls at the end of the night. Based on her reported income over a roughly four-month period, Scarlette earned around $60 per night, in comparison with a modal income across workers that was closer to $100 a night. Scarlette reported having $300 nights perhaps once every three weeks, usually generated from one or perhaps two good customers who were attracted to her demeanor and performance style. This is not an unusual pattern for women who will not allow any physical contact between themselves and customers, but who are adept performers both on stage and off.

B-drinking relies on cooperation between workers who are selling different performances. Part of the performance of sex work is about appearance. It is in the interest of a nightclub owner that the variety of services demanded by potential customers be available in his or her club. Managers are hired in part to mediate risk, as it is not always in those club owners' financial interests to be aware of when and by whom services under demand are being provided.

COMMUNITY: "UNITED IN OUR OUTSIDER STATUS"

In chapter 1, I talked about the different kinds of drinking that occur in the clubs and the importance of everyday rituals like the shared Mind Eraser drink at the start of a shift. I also pointed out that research suggests that women consuming alcohol as social activities in North American culture tend to do so with the avoidance of risk as a primary organizing factor. The community that was built among women that worked in the 1990s continues to be a strong enough one that I could follow-up on women, most of which I knew by stage name only, as many as 23 years after the last time I had seen them. This speaks to the intensity of community ties built among women who worked together as B-drinkers. I argue that this sense of community originates in strategies of risk reduction necessary for survival and is further solidified by rituals that are a part of the psychological shift one must make to code switch in and out of the B-drinking persona while in the dressing room, community ties that are firmly cemented through economic cooperation in an ambivalent occupation that is both extraordinary and illegal.

Almost every participant in my interviews spoke of the intense ties that were built among coworkers in the environment of the club where they worked. This was sometimes starkly contrasted with other clubs, in a sort of competitive stance about the relative bonding of different cliques of women. The feeling of community among girls that had faced the grind of daily life in the clubs led to cooperation that gave workers group power in otherwise vulnerable positions. The quotes below come from four different participants in my interviews in 2013:

> It was, like, a community, it was like a family. Everybody pretty much watched out for one another.

> We were united in our outsider status. We embraced that, but not the status of being a whore.

> We were protective. Like we felt like a family. That was the appeal. If you were to leave [club name], you always went back, for that sense of acceptance, protection. At the same time you are being exploited.

> It's very strong, perhaps one of the strongest communities I have ever been part of in my life. I think it's something similar, like a similar thing happens when people join gangs.

Acting like a Family: Community Support within the Club

It was a dull night at the club, with a trickle of customers coming in the door and the two waitresses immediately pouncing, competing for the limited tips for service. The manager was probably going to send one of the waitresses home because they were starting to argue. I took an empty guest-check pad from the bar and started recording the different songs the girls were selecting from the jukebox, thinking that it might be interesting to look at the lyrics as a topic. Part of an ongoing collection of lists, the guest check was doomed to be swallowed by the waters of Hurricane Katrina years later.

A couple of men walked into the club together and thwarted the waitresses by walking straight to the bar and sitting on the high-backed, black leather bar stools. They ordered their five-dollar beers directly from Max the bartender and paid him. Several dancers, hoping to score a B-drink or more, casually approached the bar for a glass of water or a match for a cigarette. Some of them were chatting amiably with Max, who was cracking jokes as usual. One dancer started chatting with the two customers, asking them how they liked the city.

"It's ok I guess except all of the faggots," said one of the men, loudly enough for me to hear from a chair across the aisle from the bar. Some of the girls moved away, some ignored the comment and kept flirting, and one casually walked down to very end of the bar to sit with Max. Max was gay, and both his brightly colored clothes and his style of conversation were performative of his sexual identity. The acidity of the remark had probably begun brewing when the tourists were farther down Bourbon Street, where there is a collection of gay bars, but the comment was also directed at Max. His tip jar was just as low as everyone else's on that slow night, and he was pushed over an emotional edge by the comment.

Ta'Mya had disappeared to the dressing room after hearing the comment and now was reappearing with her makeup freshened and a long, sequined gown, long gloves, and a feather boa. This was before she had become "Mama" and was still an exotic dancer and a B-drinker of remarkable prowess. Her costumes were always ornate and fitted perfectly. Ta'Mya was a performance on and off the stage, inside and outside the bar.

Ta'Mya minced over to the bar, moving in a cloud of feathers and perfume. Her clear target was the customer who had made the homophobic comment, and all the dancers glanced at each other wondering what was going to happen. "You *Big Thing*, what are *you* doing in my bar?" Ta'Mya drawled as she lightly rubbed one long, acrylic nail along the exposed shoulder under his tank top. She oozed sexuality, pursing her lips and flipping her long hair extensions and her red feather boa. It was only a few minutes before the customer was buying her drinks at the bar. "I only drink the big ones," she explained as he paid for her cocktail, "I'm like a Texan, I like everything big. Where are you from? You like 'em big?" Under the bar, she reached over and grabbed the front of his jeans. She sizzled and smiled as she downed a triple cocktail, a large wine glass packed with ice and champagne for which he had paid twenty-seven dollars.

It didn't take long for Ta'Mya to convince the customer (much to his friend's delight) that they should retire to the VIP room for more privacy. At twenty-seven dollars a drink, she cooed, why not get the bottle? By now, all the women were quietly laughing among themselves, watching the scene between the drunk man and Ta'Mya unfold. She stood up, over six feet tall in heels, strutting to the champagne room carrying the bottle herself and throwing exaggerated kisses to all of us over her shoulder. The customer sheepishly paid Max without tipping, and Max icily said the waitress would bring glasses back.

The waitress went back with glasses and the champagne bucket full of ice and asked for the expected twenty-dollar tip. Ta'Mya reached around the customer, burying his face in the cleavage of her gown and whispering in his ear while simultaneously pouring part of the bottle of champagne directly into the ice bucket and righting the bottle before letting him up for air. He paid the waitress her twenty-dollar tip, and the waitress promised him no one would bother the couple any further until he was finished with his time. They were in the very corner of the champagne room, visible to the rest of us sitting up at the bar in the front, carefully stealing glances in the mirrors.

We could see the scene unfolding, Ta'Mya standing up, preening and posing, and sometimes dancing for a few minutes between the customer's legs. One of the girls got on stage and started a song with lyrics suggestive of hustling money, and the girls at the bar giggled. Ta'Mya took a dollar from her customer, shimmied up to the bar and gave it to the dancer on stage, transferring the money from her mouth to the dancer's mouth. When she returned to her table we could see the couple making toasts and drinking from each other's glasses. All the while, laughing about his mate's good fortune, this VIP customer's companion stayed sitting at the bar. He chatted with the other girls, watching the dancers and buying small, seven-dollar cocktails for a few of them at a time.

Perhaps fifteen minutes later, the customer came storming out of the VIP room and headed straight out the door. He looked startled and disgusted, and he didn't even wait for his companion who followed him out the door. "What's WRONG Baby?" Ta'Mya called after him, laughing in a deep, throaty, and boisterous way. "Don't you LIKE me anymore?"

The bar exploded into laughter as the girls pried from Ta'Mya the details of her story. She had taken eighty dollars in additional tips while in the VIP room, for dancing, for kissing, and for grinding her buttocks against his groin during a private dance. She had flashed her breasts and let him touch them. At the end of her performance, she had allowed the persistent customer to finally get his hands down her thong underwear. Pressing his hand up, he had found her penis tucked between her legs.

Why did Ta'Mya engage in this way? This was the only time, over years of interviews and friendship, that I had ever heard of her exposing her genitals. She always presented herself as feminine, at work or outside. Indeed, it was only months after meeting her and having regular interactions that she told me she was transgender (she thought it was funny I did not know). I want to be clear that this was an extreme measure. Such extreme measures to protect in-group members was typical of the tight relationships built between performers in the clubs of Bourbon Street.

RACE AND SEXUALITY

As Dewey and Tonia St. Germain (2014) point out, the street-based sex workers that the New Orleans police have targeted for easy arrest disproportionately include poor, black women. In the nightclubs frequented by tourists, the women employed were disproportionately poor or working-class white women. In each case, I found at least one or two black dancers and B-drinkers employed by clubs. Variety was the key to a good lineup of dancers on any given night, and clubs ran with anywhere between 5 and 15 women depending on hotel occupancy rates and expected foot traffic around the French Quarter. That is, from the perspective of management, I was told race was like bra size or hair color—just one way to provide the right balance of variation.

African American women who were employed by the nightclubs earned incomes at least equal to white women. In the 1990s, it was rare to see any other ethnicity represented in nightclubs at all. Post-Katrina employment practices are surely different due to overall shifts in the demographics of the city, which include increasing numbers of poor Latinas who could potentially be competing for such work.

In my initial report on dancers (1993), I suggested that the strong bonding between women and the need for teamwork in the clubs seemed to be more influential over cooperation than racist exclusion. A favorite photo in my field notes is of a racially mixed group of girls sharing a booth for a portrait-style photo (never meant for publication, and only one of the women is still in my longitudinal sample of interviewees). My observations, and images like this one, had always made me feel that race was just one of the commoditized performances of identity found among the group. In a still racist America, there were creative women on turn-of-the-century Bourbon Street who commoditized on New Orleans' history as a place to find beautiful light-skinned women of color, well documented as holding a special place in New Orleans history by Alicia Long (2004). Just as Blaze Starr had her 1990s imitators, so did Josephine Baker—from her sequined gowns to her famous banana skirt performances in Europe in 1926. One mixed-race performer used a grass skirt on stage in imitation of Josephine.

Variety in appearance, variety in speaking skills, and variety in ethnicity—all are distinctions that the widget/stripper/B-girls manipulated. Their performances relied strongly on the tradition of "camp," here defined as the self-conscious parody of sexualities. Through exaggerated performance of acts portrayed in popular erotic art and pornography, creative and intelligent women fashioned niches for themselves in the market of desire.

Although I knew many more African American women in the 1990s, my follow-up interviews in 2012–2015 mainly included white women. This is a weakness of my longitudinal data; in reaching out to women for my most contemporary interviews, mostly white participants agreed, leaving my sample skewed by ethnicity. Opal is an exception. It was my interview with Opal that revealed to me the ways that African American women were less able to construct their own performances and images than did other dancers. While most dancers create their hustle, Opal reported being pressured into a role she disliked. For Opal, customers' perceptions of her ethnicity dominated her experiences B-drinking, and so she felt she had less agency in creating a performance that she identified with and that was still in demand by tourists as consumers.

Opal's responses to questions about race in the nightclubs were gloomy. In the 1990s, she was a very young dancer who preferred to wear gothic fashion both in and out of the club. Her experience in creating her stage persona was one of frustration, as she resented feeling that she was never in control of her dancer identity. Rather than being free to create a dancer persona, or a series of them, she felt that her racial identity was fetishized by customers to the point that no other aspect of her performance was visible.

> Race played a big part as far as money and being perceived as desirable. I started out at [specific club] as a virgin too . . . Men would say I was "exotic" but never wanted to spend time with me. I got the most "time" surprisingly from older white males—who worshipped me. I thought it was strange, growing up in the South. I made the most money from guys 60 and older, conservative types who would always say, 'you're a beautiful black woman.' Haha.
>
> Local and tourists, I always got the, 'so what *are* you?' question. Being biracial wasn't easy. Men couldn't quite place me into a stereotype so that messed up their fantasy. I wanted to be my goth self. I never wanted to play a role 'cause I wasn't good at being someone else.
>
> Well after I left dancing, I was a bartender in an upscale hotel. Same men—different place. I was shocked. I thought I left that world behind, but no. The only difference was I wasn't a stripper. The same rhetoric as before but I had no bouncer to kick them out.
>
> It was confusing, I was completely covered and worked for a global company—health insurance, paid vacation, and a new start at life. I wasn't working seedy bars anymore but at 20 you can be naïve. I guess it was the learning experience I needed.—Opal, personal interview, 2015

HOW TO B-DRINK

Most B-drinking happens in the context of breaks between explicitly sexualized striptease/pole-dancing sets. Mingling with guests capitalizes on the tourists' urge to engage with local women. The least visible (to me) B-girls reported working as bartenders outside of the strip clubs, like Opal's experience in an upscale hotel bar. Similar to striptease artists, bartenders used B-drinking as a way to increase their incomes and the income of bars throughout the touristscape of the city. The practice added significantly to liquor

Figure 5.2. Word Frequency Cloud of Interview Content on 'Stories about B-drinking,' created with Nvivo. Created by author

sales and in the strip-club context insured the owner of the club that at least a small percentage of money taken from customers would end up in his cash register. How much of that alcohol ends up in the dancers' bodies also varies a great deal by personal strategy and correlates very little with how many drinks are purchased for the girl in an afternoon or evening.

The B-girls appearance soliciting on "the floor" of the nightclub represents a much longer period of her work time than the time she appears on stage. In a small nightclub, she may be one of between 6 and 12 performers that evening. Much less frequently, waitresses B-drink, and very rarely a woman is allowed by management to B-drink without doing either. This happened when women were off the clock but stopping in on a very busy night (such as after a football game) and were invited specifically by the customer. Similar occasions occurred when retired dancers visited the club and were offered drinks. In both such occasions the customer himself would have solicited the woman to accept a drink—women who were working would have been enraged by a visiting non-employee actively soliciting drinks and thus stealing potential customers.

The B-girl dresses herself compellingly, constructs an identity for herself with a seductive name, and crafts her ability to flirt and chat in exchange for drinks. She may giggle or sizzle, depending upon the niche in the market that she hopes to tap. In my structured interviews, as well as in a blind survey completed online, women who reported on their experiences B-drinking did so with neutral or positive comments.

> Well it was kind of the job! So if you wanted a job, in most of the clubs you had to engage in B-drinking, it wasn't like—I never felt coerced into doing it. I was pissed off because I think that the wage scale was really . . . unfair. You know so for instance if I remember correctly, we had to first start out selling 7 champagne cocktails, that were priced at $7 apiece, in order to meet our quota, and once we made our quota then we got a certain, um, hourly payment. Which, that would be like $35 per night. And from that $35 and our tips, we would have to tip the doorman, tip the bartender, which in my experience was always male. And, at any rate, in order to work you had to B-drink. And it was . . . I know that it was illegal, but I never felt, like worried about it, worried about doing it. There was never fear of going to jail for B-drinking, or anything like that. It was just something we did, it was part of the work culture, and part of the . . . the pay scheme, if you will.—Sadie, 2013

As soon as a customer walks into the club, a waitress meets him and begins the process of guiding him through a scripted process of finding him a seat and getting his order; while none of the three clubs observed charged a cover charge, all required the purchase of a minimum number of drinks. Once a customer is seated, women who are working the floor between sets are free to approach him, sit down, and strike up a conversation that will hopefully lead to a drink, a private dance, or both.

When business is slow, waitresses hang around doorways and engage passing tourists in conversation. It is easy to spot tourists who are on a pub crawl through the French Quarter, with their full glasses of colorful cocktails and out-of-season Mardi Gras beads. Doormen and bartenders, like some dancers, keep informed about what sorts of conventions are in town, which can help to sort through larger crowds of tourists to target the biggest fish. Knowledge of the crowd helps a doorman understand what register of speech, what style of joke, or even what style of dress might bring curious passersby in for at least one round of drinks. Waitresses sometimes assist the doorman and through chitchat can start the process of finding the right B-girl for the customer(s).

Within the environment of the strip club, the approach of B-drinking provides an interaction that seems normal to most customers. As I have discussed, it is the norm in American bars for masculine people to buy drinks for feminine individuals, making the literal and measured commodification of the B-girls' time nearly invisible to the customers. The difference between this and a normative American script of seduction through intoxication is that the performance of a B-drinker is commoditized emotional labor, and her power of seduction translates to a wage. Although this inverts gendered relations of power, it is accepted as part of the range of inversions and marginalities that dominate the tourist's landscape of the French Quarter.

B-drinkers do their best to earn their money talking rather than in contact. They may specialize in whispering sexual talk, work that is very much parallel to that of a phone sex operator. One young woman in my sample earned quite a sum of money talking to customers about the welding classes she was taking at a community college. The B-drinker talks to men in nightclubs about all the same topics women talk to men about in bars in normal American bar culture; that is, about everything imaginable.

In describing gendered drinking patterns, Jaqueline Ferris (1997, 675) discusses her interviewing process, explaining that her semi-scheduled interview format began with "throw-away (warm-up) questions . . . followed by the core, or essential, questions." She defends her interviewing methods by referring to Bruce Berg's (1989) text on qualitative interviewing, thus legitimizing her analysis by framing it as based in a history of social-science research methods. One participant in my interviews also discussed the use of such "throw-away" questions as a part of her everyday hustle for B-drinks in the bar, with the essential question of whether he would buy her a drink delayed. In describing her strategies as a B-drinker, Jennifer shared the following story:

> I've also done car sales. The car sales job . . . we would have weekly sales meetings. And you really learn the art of closing the deal. And some of the things that I was already doing intuitively from trial and error, or watching other people . . . I mean, they do studies; they put a lot of time and effort into how to sell something.
>
> Like one of the things is you get three yesses out of them before you ask them for a commitment to what you want. So, you say, "Are you having a good time?" "Yes." You know? "Do you like the city?" "Yes." Because it gets them in the momentum. That hustle, that salesmanship that is the most important part of the job. When I first started dancing I was young and cute and I didn't make *any* money. I learned how to hustle from older women, obese women, women missing teeth. —Jennifer, interview, 2013

Individual girls may specialize in hustling men into believing they are potential sexual partners, when in fact they are earning a living keeping the client engaged in such conversation long enough to exhaust his spending budget for an evening of his visit to the city. On the back couches of the VIP rooms at 2 am in any number of clubs up and down the strip a conference attendee can find young women like Amber.

When I met Amber in the mid-1990s she was a twenty-three-year-old woman wearing a short dress over a black thong and a pair of knee-high, tightly laced 5-inch platform go-go boots. For $150 for half an hour, including some drinks, Amber and a customer could enter the VIP area. It is in this environment that we find our 1990s "Lothario" or "swain." Finding he has already paid for the room and the absence of further harassment, he is certainly willing to pay a fair price for whatever service he is seeking. In some cases, the VIP room then became the site of perhaps more obvious sex work. Amber was one of the girls willing to give hand jobs in the VIP room, but only for men who were purchasing a second bottle of champagne. This means a minimum of $300 in champagne, $5–8 for the customer's first drink, and $40 in tips ($20 for each bottle) delivered by the waitress. Most of the girls agreed that for $345, a discreet hand job was not stigmatizing behavior. After all, there were several girls in The Den who were willing to leave with a customer for sex in his hotel room for around $300.

Still, in most cases and depending on the style of hustle that the girl has developed, the B-drinker's customer will not receive actual sex. Instead, like many sexualized encounters in topless clubs around the United States, he will be engaged in verbal and nonverbal (dance) messages that are highly sexualized in nature; all B-drinking requires the performer to determine and provide the emotional context desired by her customer as a critical component of her work.

Much of the women's work resembled most closely that of escorts, earning a slow, steady income drink by drink to sit with men while they watched the show. It was critical that they cooperate with one another, encouraging men they were sitting with to tip stage dancers or buy drinks for their friends, particularly if a man was identified as a slow customer. Such customers sit as long as possible, buying small drinks and never bottles, and asking the waitress for single dollars for change.

During my interviews in 2013, I learned that at least one club's management in the 1990s had also engaged in a practice called "buy outs," although I never observed such a case (reported to have been a more common practice in the 1980s). Panther and Sadie both reported that this practice was one way in which management was openly involved in out-call prostitution.

> Back in the time when there was [nightclub], there was also [another nightclub], which was like the lowest of the low. I mean I worked there one night and then that night they had buy outs . . . which meant that someone could pay a certain amount of money, like three hundred bucks, and get a limo (included), and they get to take them [the dancers] OUT of the club. —Panther, interview, 2013

WORKING TOGETHER AND BUILDING COMMUNITY

Several participants in my structured interviews talked about the importance of cooperation with other workers in the club in order to make their incomes.

Because of this, B-drinkers depend on waitresses and other dancers to act as intermediaries in selling drinks:

> I wasn't even good at making quotas. It's just because I didn't like . . . selling myself. I didn't feel comfortable with it. You know if I was drunk, I was a little better at it but . . . I was better at hustling other people than I was at hustling myself.—Brie, interview, 2014

> I wasn't very good at hustling for myself, because I just never felt comfortable with that, I felt like that was degrading. I didn't want to sell myself, I wasn't comfortable with that. I had no problem selling other people, as a waitress, at one point.—Panther, interview, 2012

> I started working with Cassie because I was really good at getting people up to VIP and she was good at stringing them along and getting second and third bottles. And I realized that two stupid girls are still smarter than one smart guy.—Chiante, interview, 2014

> I liked the soft hustle, and I . . . you know . . . the waitress would come and ask if I, if he wanted to buy me a drink and I'm like "I'd love to stay here, but they, you know, I can't. . . . If you don't buy me one I can't." So it was kind of a safe way to interact with the men, and more organic, I had thought than just going around and ask. And every time I did leave the club to work somewhere where there wasn't B-drinking I never, I never fared . . . as well? I just like, I hated it.—Cloudy, interview, 2013

Sometimes, a B-drinker can develop rapport with a customer during their shared interaction, which contrasts to the disruption by the waitress asking for drink orders. Here, the B-drinker can deflect the customer's sticker-shock about the price of drinks off herself and onto the club, playing the oppressed woman. This also can increase her tips, as Amber reported.

> Sometimes you would say to customers, sort of in a pissed off way, like "can you believe this bullshit?" right? To explain that this is how I have to make money. Some of them say "look, just hang out and talk to me, and don't get a drink, and I'll give you the money," and like, well, as long as . . . . As much as I can do it . . . so it gives them the hope that the whole thing is organic, in a way [laughing].—Amber, interview, 2013

Women also sell drinks for other women by acting the part of management. One type of response for the waitress in a situation like the one above is to take an aggressive stance on the rules of the interaction. Particularly for tourists who have never before but are suddenly bemused to find themselves in the VIP room, the waitress provides the structure of the interaction. The customer is agreeing to a sort of suspension of disbelief in his interaction with the dancer with whom he is sharing an expensive bottle of champagne because this is what he has paid for. The waitress, or bartender at the bar, can be more removed in explaining exactly how things work.

> On the floor, you were selling. You were using sex to sell, for the girls. Although I'd never do that. I'd always use pure intimidation. Like, being totally crass, and like mean, to sell cocktails. I don't know. I mean, to me it seems like more protective sort of thing. The men were there, but in the sort of foreground whereas the women are on the floor, or on the stage.—Sky, interview, 2013

For the most part, women who were waitresses or bartenders in nightclubs where B-drinking was taking place were discouraged from accepting B-drinks. Because of the structural role of the waitstaff as separate from the entertainment staff, waitresses who accepted B-drinks were a recurring source of conflict between otherwise friendly groups of employees, particularly during slow shifts. The waitstaff must meet every client at the door to enforce the drink minimum on which the nightclub depends. This leads to dancers feeling shortchanged when waitresses accepted drinks and thus commissions. Dancers felt it was unfair when waitresses earned drink money in addition to the tips they would have already received for delivering the customers' own drinks. When waitresses engaged in too much B-drinking, dancers sometimes responded by pilfering the serving tray and seating some newly arrived customers, collecting the tips on at least one round of drinks. Management tends to stand back in such situations and let the disagreement play itself out, as long as no customer waits for service. I was told that a good waitress, while getting a customer's own drink order, should have been busy discreetly assessing the contents of his wallet, the brand of his watch and shoes, and the kinds of girls he is watching. She doesn't have time to have a drink.

"UNITED IN OUR OUTSIDER STATUS": ACTS OF RESISTANCE AND PERSONAL AGENCY

One of the most common topics women reminisced about in my 2012–2016 interviews was a feeling of being in control in deciding what sorts of work and conditions were acceptable to them. Sometimes control was framed as being in a binary to concerns of management and expressed through acts of resistance to club or management demands. Like the dancer who resists being rushed out of the dressing room by appearing on stage half-costumed, women who worked on Bourbon Street told me many stories reflecting their feelings of power *within* the club. There is an almost constant push back to any control on the part of club management that the contracted employees felt was unfair. Bourbon Street performers are nearly as protective of their right to creative freedom as tenured academics.

> We were very resistant, we engaged in all these resistant practices, like either not adhering to the laws. Or just the act of B-drinking in itself was a resistant practice against the laws. We just DID it, it was part of the culture. We didn't care. So . . . I know for myself that although I was engaging in this practice that to me was [deep sigh] degrading in some senses, I did a lot of things that were sort of in the face of that. In the face of degrading practices.—Amber, interview, 2013

> He (the manager) was always so full of shit though, he would tell you what to do but then you would not do it, and he wouldn't do anything. It was funny.—Angel, interview, 2013

> You could play [on the juke box] whatever you wanted to, and even though [the manager] would bitch about it, he would be like "you can't PLAY [Nina Simon's version of "Strange Fruit"] while stripping because it's about lynching, and people don't find that sexy," you would do it anyways. Because I didn't want to fit the stereotype of a dancer.—Sky, interview, 2014

The economic need to cooperate with other women, such as other dancers or waitresses, paired with the conflicts between male management and female employees led to a strong sense of comradery among the women in my sample of B-girls. Using their power as a group, a great deal of the "rules" of any particular club were constructed from within the dressing room. It was critical, for example, that the women agreed to a reasonably similar range of prices for the kinds of sex work that occurred within the club. A newcomer who undercut what were considered fair prices by other girls would either be run off or would have to quickly change her pricing expectations. While having girls who charged higher prices within the club was less problematic, one cheap worker pulled down the perceived value of work of all the women working the club.

> There were some people who deviated from, I guess, the community rules. We used to be like "oh look at what she's doing, she's like letting that customer stick his finger up her cooch, for one cocktail!" Because we tried to distinguish ourselves, at least in my experience, from those who did really nasty shit—even if sometimes we did and didn't say anything about it. You know?—Candy, interview, 2014

Candy is not concerned about the nature of contact between the hypothetical coworker and the customer. She in fact acknowledges that similar behavior would be reasonable, even expected, if the customer were paying a fair price. In an independently published "zine," Janet expresses a similar sentiment concerning the "code of ethics" created and enforced by the group concerning pricing.

> Competition breeds malice and envy. It's capitalism on a raw, personal level. Most of the women I've worked with won't cut your throat for a dollar. There's a code of ethics these dancers live by that strikes a balance between making money and co-existing civilly with the other dancers. Enforcement of the code is usually worked out within the group. Sometimes with a reprimand, sometimes, for a repeat offender, with a fist.—"Janet." *Rocket Queen* (independently published "zine," catalogued at the Newcomb Archives for Research on Women)

B-girls calculate what they are earning for each interaction and surreptitiously compare that to other women in the club. They also are often critical of the percentages of each customer's money that is claimed by the club for their work, as well as by bartenders and other male employees who claim a percentage of their cash at the end of each night in tips.

> There is like a mathematical formulation in your mind of, like how much of the bottle money am I gonna get, plus what tip am I gonna get, right? And the bottles allow [the owner] to keep the lights on, whatever. But from the girls' perspective we would have made more money with table dancing. Like in San Francisco, it's no drinking and all nude lap dances, but you had to pay the house $240 for the shift.—Diamond, 2014, interview

In perhaps their most powerful act, B-drinkers and dancers can stage a walkout. Because they are independent contractors, particularly during the busier tourist seasons it is relatively easy to leave the employ of one club and find a job at another. Enabled by the close friendships formed in the dressing room, cliques of women can plan a shift from one club to another as a group. If the management's perceived transgression is offensive enough, if the complaint is robust enough, or if the season is slow and boring enough that may convince a large group of the women employed by a specific club to drift together.

> There was another place, it was called [club], one of those tiny straight back places. Because there were times when—both of the times when I went to those places were times when I was pissed off at [manager/bartender], and so I was like "fuck you guys, I'll go somewhere else," and in both of the times it was like me and a bunch of other girls left together.—Sadie, interview, 2014

Such collective movement causes a minimum of disruption in the work and cash flow of the dancers, as by moving together they ensure the assistance of coworkers with whom they have already formed cooperative scripts both on and off stage. For management, such a move had the potential to (temporarily) shut down the abandoned club. No one is interested in drinking in a strip club without dancers. Such walkouts were one thing that would force management to spend time after hours (after 4 am, for Bourbon Street employees) visiting bars popular among B-drinkers/dancers to try to rebuild an employee base. Usually, such walkouts are temporary, and management gives in to the groups demands.

> We always went back. I think I did with other girls, and we left, but we always went back to [club]. Because the other places, there wasn't that . . . cohesive sort of community that we had developed.—Diamond, interview, 2013

> I have seen where girls have all, one time at the [club] all the girls walked out and had a strike. And that manager did mellow out after that. But I think that just the way those things are set up in the first place, is all to cater to our *rock star* personalities! Otherwise . . . it's like I have the pussy, I make the rules. Like, without us all you have is a bar with overpriced drinks.—Panther, interview, 2012

DRUGS

Another variable of club life that was mainly controlled from the dressing room was drug use. Although some of the women I interviewed in 2014 discussed their past opiate addictions with me, during my observations in the 1990s such drug use was carefully hidden. Drugs considered "party drugs" by the women, such as marijuana and cocaine, were expected features of the dressing-room space every night. Needle drug use, however, was hidden and broadly considered taboo in the clubs, in part because B-girls feared that addicted women facing withdrawal would ignore community agreements about pricing for work.

> There was a line. So, it's like recreational, fun is OK. But anything beyond that is not. And people did participate in that (heroin use), but not . . . well, so one person I knew had a really bad heroin problem and ended up working day shifts giving hand jobs. I mean I like her. She is hard to work with.

Melissa Lavin's observations in a strip club she calls Mell's Belles, which she classifies as a "low-level strip club" (2014, 136) in a northeastern city, suggest that "strippers mobilize space to avoid being seen using discredited drugs." She refers to the strip club as a "drug space" (133), which provides and normalizes access to a range of legal and illegal drugs, including alcohol, which she calls "contextually encouraged . . . in strip clubs and in larger society" (133). In contrast, Hanna (2012, 219) claims that drug checks and even drug-sniffing dogs are common in upscale clubs. Hanna argues there is no evidence that drugs are any more of a problem among dancers than other occupational groups. These two oppositional portrayals are typical of the two extremes of documentation about strip clubs.

Lavin observed dancers drinking alcohol and smoking marijuana in open spaces but hiding other drug use (2014, 137). Unlike Matthew DeMichele and Richard Tewksbury (2004, 547) who found that male bouncers enforce club rules to prevent widespread use, Lavin (136) found that some club management and bouncers sold and used drugs themselves and only sporadically acted to modify drug use by modifying the environment (such as the removal of bathroom doors, also found in clubs in my own sample).

Lavin describes an "outdoor area that is visible from the main road. . . . Club actors go into this area to smoke marijuana or cigarettes and to drink and converse outside" (141). Although I found marijuana use was also normalized within the strip clubs of Bourbon Street, such activities occurred in the farthest "back stage" area—the dressing room. Unlike Mell's Belles, the clubs on Bourbon Street do not depend on a local, working-class customer base, and the outdoor areas of Bourbon Street are carefully policed. Customers on Bourbon Street are not local men as at Mell's Belles; for the most part, they are middle-class people who can afford to vacation or professionals on an evening stroll after a conference. The scene Lavin describes of a dancer sharing drugs with customers outside Mell's Belles is unlike any drug-use patterns I have seen on Bourbon Street.

Lavin suggests that attitudes toward drug use in Mell's Belles "generally mirror those held in broader society" (2014, 142). She found "hard" drugs, to include crack, cocaine, ecstasy, and heroin, to be "most scandalized, both among strip club actors and more mainstream groups, especially when snorted or injected" (142), and she suggests that users of these "hard" drugs cover their behavior by passing as users of "less maligned" drugs. She cites the use of alcohol, marijuana, and prescription pills as carrying little stigma in such clubs. In contrast, I found that in clubs like The Den on Bourbon Street neither pills, marijuana, cocaine, nor ecstasy carried stigma within the group as long as they were not injected. Less common but seemingly no more shocking were dancers who smoked crack cocaine, although this was a drug used by more transient women who worked for only a few days. Shared use of any drug among a group in the dressing room, including crack cocaine, increased social ties among the group of contract dancers.

Lavin's work highlights the importance of considering the differences between strip-club behaviors in different cities and especially differences between clubs that serve tourists and those who serve regulars. B-girls do consume alcohol with their customers—but they do not consume other drugs in any of the front-stage areas of the club. As their job entails seducing customers to purchase them alcohol for an income, they do not freely share other intoxicants with potential customers.

Additionally, Lavin's description of regular patrons using drugs to "barter and trade them for lap dances" (145) is an unlikely scenario in a well-monitored Bourbon Street club. Women who are performing for men for drugs add nothing to the club cash register, and drug sales on the main club floor are monitored. Even when management is aware of narcotics use in the club, there is strict control of who introduces the drugs into the building. In one instance, I witnessed a two-hour standoff between a club owner and a known drug dealer. To assert his control over a particular dancer, her heroin dealer purchased $150 bottles for her and another girl and retreated ostentatiously to the VIP room. The club owner chain-smoked quietly at a table just outside the VIP room, irritated with the presence of the dealer, but allowed the "party" to continue through the purchase of three sets of bottles. As soon as the last bottle was empty, the owner physically escorted the dealer to the door, shoving him into the street. At the end of the night, the owner withheld the commission for these bottles, causing an argument that led to the involved dancer being fired. Club owners do not tolerate competition over the control of space within the boundaries of their business.

Drugs do enter the clubs in the French Quarter, but they tend to enter through the dressing room. The dressing room, as a far back-stage space, with zero access for potential customers, is an area where non-intravenous drug use was of little interest to the management. Still, as Bourbon Street performer and heroin addict Eliza Player (2010) describes in her memoir, dancers had to be careful to manage the front-of-the-house impression that the club was an inviting atmosphere for businessmen.

> I put on one of my most street worthy stripper outfits . . . I stepped out of the doors, bullshitting for a few minutes like I was hawking customers. Then, I turned to the left and darted down Bourbon to the corner. Turning on Conti, I picked up the pace as I rushed to meet the man with no less than two hundred dollars in my boot . . . I hopped in the car and quickly made the transaction. Hugging tightly to numerous little foils of my precious dope, I headed back down Dauphine towards Canal for one block until I turned on Bienville. Walking up Bienville to Bourbon, I passed several bars and fine dining restaurants. I made the block at Bourbon and was back in the club lickety-split. (Player 2010, 47)

Congruent with my own observations and interviews, Player's memoir represents drug use as common in Bourbon Street clubs. It was ignored so long as it was not obvious, did not disrupt business, and did not cause fights.

REMEMBERING THE PAST

Women in the clubs worked cooperatively to create an atmosphere that they agreed to share. These nights of tacit cooperation, even between girls who did not necessarily get along outside the clubs, has led to long-term ties. Women who worked together in the early 1990s are still in communication with one another today, sometimes in ongoing personal relationships and also using social networking websites. Today, in instances of individual crisis, community social networking sites serve as a point of contact to gather supplies for needy group members, help one another find or rent out apartments, and provide emotional support to women who otherwise may feel stigmatized by their past as B-drinkers.

I was invited to join one social networking site by the site's administrators who knew me in the mid-1990s. I participate as a member, but I also have used the site to contact women for longitudinal interviews. Figure 5.3 is based on data that was mined from this ongoing, online group that serves as a virtual meeting space for the women in my sample. The conversations that happen in this virtual meeting space are confidential, and I will not quote them. However, participants in the virtual space are aware that I am an anthropologist, and I have used the webpage to organize interviews as well as to launch an internet-based wellness survey. I also regularly donate to fundraising done for community members in need. Although it would be unethical to quote women (outside of those who have given informed consent for specific content), I do feel the group's online discussion can provide useful data about interactions and topics of shared interest within the group. With only a few exceptions, the members of this online group no longer work as B-drinkers or striptease dancers.

The most common theme of posts were videos of songs that the participants associated with their work on Bourbon Street. Often such posts reference certain time periods or individuals that they remember working with while hearing the song. This kind of post exemplifies the nostalgia the women themselves feel for their past time as a unified, working group.

Like in any online group, there is a wide variety of memes relevant to the community outside of their role as B-drinkers and simply posted as a sort of internet greeting. Posts that expressed solidarity with the group, such as "I Love ALL my sisters from The Den!" directly reflected nostalgia for time spent working together. Equally as common were posts initiating reunions large and small, including requests for help locating group members not yet connected online.

My claim that B-drinking on Bourbon Street contributes to long-term group solidarity is supported by the number of posts which served to communicate about attempts made to assist and fundraise for members of the community who were in financial distress. This included raising money for substantial health expenses or relocation in emergencies and attempting to assist group members who currently face homelessness. Figure 5.3 simply reflects the *number* of threads that were started within each of these broad themes.

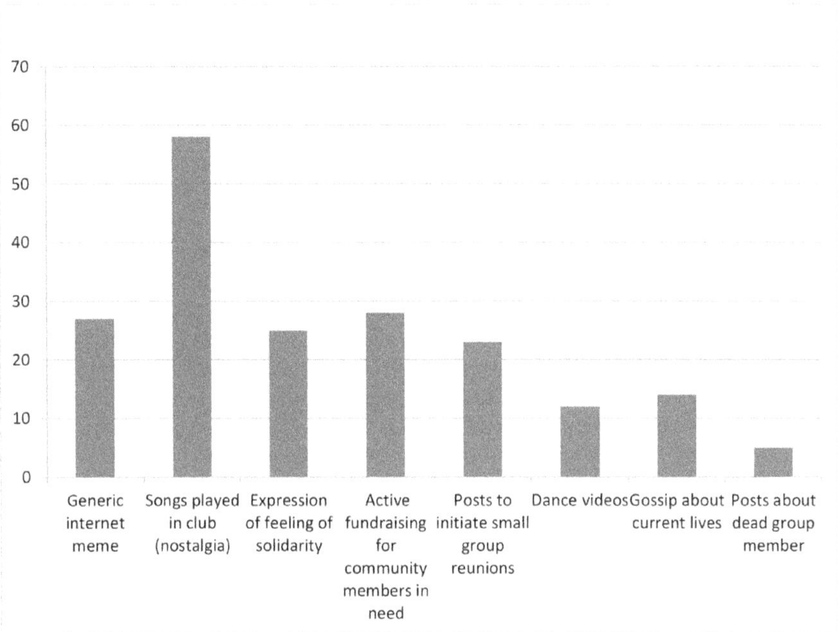

Figure 5.3. Most Frequent Themes of Threads on Social Networking Site. Created by author

Another important variable in looking at such threads is the intensity of response the topic receives from the group. Aside from there being twenty-eight separate threads concerning fundraising at different times, by far the longest threads, with the most participants and largest number of posts, fell into the thematic category of assisting group members in the contemporary moment—more than a decade since members last met. The attempts to provide basic supplies like clothing, a blanket, gift cards for food, and feminine wipes for one homeless member of the group created such a stir that she could have been coded in my notes as her own theme. Women now located across the United States checked in on these posts, as those still living in New Orleans searched beneath the overpasses of the city for this member's current camp, hoping to deliver basic supplies and much-needed words of encouragement to this lost group member. Likewise, although there were only five threads specifically addressing the deaths of members of the group, these threads were followed and commented upon by the largest number of group members.

I end this chapter with a dream about that need for cooperation, recently recounted by one of my participants and shared with the group. With her permission, I share the transcript of a video she posted in 2013 when she reconnected with other group members through the social networking site.

> I had a dream one time, and it was one of those very striking dreams those few and far between that you have that really stands out. It was a detail oriented dream, I dreamt like the whole night, like, picking my music out in back on the jukebox and dancing my two songs, um picking out which songs I was gonna dance and how I was gonna dance it. Going around asking for tips and it happened to be a very busy night at the . . . bar. And, um, everybody was making money, bottles were going off [popping noise with her mouth] corks flying across the room.
>
> And I'm sitting on a triple with a dead beat—and a waitress, who can't figure out when I *roll my eyes* when she's like "Everything okay?" and the guy says "Yep!" and I'm sitting on an empty drink, and he's trying to paw me and *she's not pushing the drinks*. And I'm trying to do the soft sell and it's not working. And I'm very pissed off and she walks off but I'm just trying to maintain my cool and when he goes to paw me I just lose it and I go to hit him and (aggh!) and I run to the dressing room and I'm kicking lockers. And there's no one in the dressing room because, well, everybody is out there on the floor making money.
>
> And who comes in but Stephanos, and I'm kicking and he's like "What's the matter?" and I start venting "I'll be damned if I'm sitting with this guy for half an hour with an empty drink and the waitress does not get eye language, body language and the eye signals I am making to her to, like, push the drink or end the time. And everybody's making money and I'm wasting my time, and getting pawed up." And Stephanos doesn't miss a beat. He says, "You know what that man does for a living? He's a gynecologist. He makes his living off of women's pussy. And it's your job, on behalf of all the pussy, to get that money back."
>
> That was my dream. It was like a joke with a punch line.
>
> I always said that to people when I talked about our era that it was something beyond, like, another stripping group of girls. It was really like a sisterhood, it was really something unique about what we experienced and the time that we experienced it. And this reunion is proof of that, and it's kind of cool I think because I haven't had a high school reunion and I feel like this is kind of the equivalent but only better.
>
> (April, 2013, recorded at the flyway alongside the Mississippi River behind Audubon Zoo.)

NOTE

1. Immediately following my own evacuation from New Orleans due to the flooding associated with Hurricane Katrina, I volunteered for several weeks at a shelter in a neighboring state. As part of my work, I directed the creation of a drop-in daycare that met the requirements of the state. Soon after opening the daycare, other people doing organizational work accused one of the young evacuees of working in a local strip club. Rather than seeing this as an admirable attempt to help her destitute family, the shelter workers felt she and her family should be forced to leave the shelter. I mention this story because it attests to two things: first,

that work in strip clubs allows women access to a living wage, with immediate access to that income. Second, it attests to the degree of stigma women incur for working in strip clubs, even in times of critical need.

Chapter Six

Dangers

To balance my informants' stories of group and personal agency, this chapter focuses on the difficulties and dangers that are a part of being a B-drinker. Much of this revolves around the stigma attached to working in strip clubs on Bourbon Street. Erving Goffman (1963) uses the phrase "spoiled identity" in discussing the notion of stigma as a social construct. This spoiled identity is a form of social control, congruent with my assertion that the continued illegality of B-drinking even with very few arrests strengthens social power over otherwise legally employed service workers. B-drinking and other work within clubs is controlled using stigma, here buttressed by a historically gendered law. Goffman's writing predicts the development of association groups for support among stigmatized populations of various kinds (including prostitutes). The online communities and long-term relationships that facilitated my longitudinal interviews corroborate this prediction. In the case of B-drinkers, my interview data supports the hypothesis that social stigma leads to social control and permits violence.

Certainly workplace violence occurs within some clubs, as was the case of a performer who was caught on camera "dirty mixing," or engaging in a sexual act (oral/genital contact) while doing a lap dance. This event occurred early during an evening shift when there were few customers or B-drinkers on the floor and many women in the dressing room. The owner of the club stormed out of the back office and into this farthest corner of the champagne room, grabbing the woman by her hair and dragging her back into the office, right in front of me. The customer left quickly, and the performer emerged from the office about half an hour later to continue her shift. Such contact between management and contract workers caused gossip, but only briefly. It was viewed as an expected response to her lack of caution in dealing with surveillance. No statistics on the frequency of such incidents are possible because the victim of assault was also potentially under threat of arrest and could not report the incident.

The conditions of the dressing rooms can also be viewed as contributing to structural violence; changing clothes and eating meals in a space with a lack of climate control and pest control, and the dangers of dripping ceilings and exposed wiring in old buildings certainly increase the chances of accidents or illness. Although the front of the club must attract the tourist, repairs not visible to customers were of little priority. Knowing they are a stigmatized population, women in my sample did not question the poor working conditions found in the strip clubs and other bars where they were employed. There is a distinct demarcation between the spaces inhabited by female employees (such as the dressing room and ladies' toilet) and spaces used by male management and customers. Unsafe and unclean working environments seem to be not only a matter of economic thrift but also function to signal to the female employees their low status. Such manipulative conditions work to reinforce notions of low worth, contributing to the tacit control of employees, and are reported of strip clubs elsewhere (Kempadoo 1998,186: Sundahl 1998,178).

The dressing room is a place where "dirty work" (Ashforth et al., 2017) is introduced to the new performer through "divestiture . . . the stripping away of newcomers' incoming identity and attributes so as to reconstruct them in the occupation's or organization's image" (Ashforth et al., 2017, 1268, Van Maanen and Schein 1979). Ashforth et al. (2017) use "exotic entertainers" as an example of dirty workers and point out that some "dirty work occupations . . . disproportionately utilize divestiture" as a strategy to socialize workers into such stigmatized occupations. Divestiture is a tactic that teaches individuals to "think and act in ways that are unfamiliar to them prior to entering their training . . . used in organizations such as religious cults, radical political groups, and organized crime families" (Jex and Britt 2014, 88). Whether women arrived at the club seeking money to support addictions or tuition, their outside identities were irrelevant to their work and blunted within the dressing room.

Issues of discrimination against and dangers posed to B-drinkers outside the club must also be addressed here. Whether being refused a lease on an apartment (interview, 1994; also reported by Hanna, 2012, 229) or declined while attempting to open a banking account (interview, 1995), the marginalized status of women who work on Bourbon Street was reported to create multiple barriers to their everyday lives. This combination of stigma and rejection in daily life, paired with hard economic power in an industry that benefits from the work of B-drinking, creates a double bind for the worker. B-drinkers can be found wavering between pride and shame in their employment. Sudden emotional breakdowns, occasional needs for assistance in negotiating the "straight world," and moments of individual women questioning their agency at work or home were all parts of everyday life during my observations of B-drinkers in strip clubs. Though life in a career path that includes B-drinking is difficult, women in the 2010s that retrospectively reflected on a B-drinking past generally believed that their economic hardships would have been more damaging than this stigma if they had not been involved in the industry.

SOCIAL REJECTION BY ROLE MODELS

Two observations of the interactions between popular culture entertainers and women working as dancers and B-drinkers in the 1990s illustrate their continued role as stigmatized public women. New Orleans has perhaps more than its fair share of famous and quasi-famous inhabitants—the city both produces brilliant artists and attracts them as temporary or permanent residents. The influence of such persons on the tourism market is strong—for example, being able to add information about the location of famous homes adds to the tour guide's repertoire directly. Additionally, artists in residence have indirect effects on how those who sell tours of the city construct those representations.

Briefly mentioned in my preface, in 1993 Anne Rice purchased the historic nineteenth-century building that had previously served as the St. Elizabeth's Orphanage at 1314 Napoleon Avenue as one of her New Orleans' homes. In 1994 her famous novel *Interview with the Vampire* (1976) was released as a movie, produced by Geffen Pictures (Jordan, 1994). A detailed description of the Second Empire design of the three-building St. Elizabeth's fills several pages of Rice's *Memnoch the Devil*, released in 1995 as the fifth in her Vampire Chronicle series. The early 1990s were also a time when a wide range of vampire stories were added to the "Ghost Tours" given throughout the French Quarter. Houses located between two already existing tour guide stops were selected, and individual guides busied themselves composing stories of college coeds who had been found inside, hanging upside down and drained of their blood. Local celebrities, Anne Rice in this instance, are contemporary powerful voices in the construction of the imagined New Orleans and create everyday spectacles when they choose to appear as an "everyman" in the French Quarter.

My first experience observing a popular artist who contributed a stigmatizing voice within the strip clubs was a musician. Perhaps more so than other residents, it is rather difficult to impress a bar girl in the French Quarter with local fame—it is not unusual for musicians performing at festivals to visit Bourbon Street clubs. However, this was an internationally recognized musician, whose dark and often erotically themed music was on the jukebox in this club and regularly played by the youngest dancers. He was in his late twenties, handsome and edgy. He was the sort of temporary inhabitant of the Crescent City that locals are excited to report seeing in the supermarket. The fellow came into the club, was recognized and approached.

The girl who was next on the dance list played one of his most popular tracks as her first song as she strutted out on stage, smiling and excited. He was furious. He was rude. He told her, and the girls around him, that that was *his* song. "I am [artist's name]!" he hissed, demanding the song be turned off. Insisting that the song was a reflection of his personal pain, he disgustedly reported it was not "meant for a stripper." The dancer was indignant, and returned his statement—"Yeah? Well I am *Gigi*. And I can dance to anything on the jukebox that I fucking want to." Although her response was quick and assertive, the confrontation was still hurtful to the young woman involved. Many years later, chatting on one social networking site, the incident was the subject of a lengthy discussion and much reminiscing. Gigi was respected within the group, at least in retrospect, for (in the eyes of her peers) putting this celebrity in his place. At least within the boundaries of the club, where Gigi was supported by a group of other dancers, she was able to maintain dignity under attack while on stage.

The second example concerns the behavior of an actor who had, a few years before, starred in a film in New Orleans. A club on Bourbon Street was used as a set for a short clip in the film, in which he is the hero and the club is meant to set a scene of vice unrelated to sex (building another narrative that features the secondary-effects myth of strip clubs). During one late shift while I was in that club, it was a slow evening and I was able to hang around the bar chatting with a group of performers. The doorman was also a friend of mine and called me up to the front.

He was also chatting with someone in the doorway, and I glanced up and down the street. He laughed, asking me if I didn't recognize the man he was talking to. For a few minutes I was overwhelmed, processing the difference between the screen and real appearance of the man. As he started to walk away, I realized I had missed a moment and ran after him into the middle of Bourbon Street to ask for a photo with him. He refused, mumbling something about how his (also famous) wife would probably not like that. I realized at that moment what I looked like and how such a photo might be perceived by the media. It was a moment of denunciation that impressed upon me the rejection felt by these service-industry workers, even inflicted by those whose careers were built on narratives and representations of Bourbon Street nightclubs! Both individuals used the space of Bourbon Street as a place to develop their own careers and to be entertained, but both rejected contact with the legitimate inhabitants of the spaces they purport to represent.

PHYSICAL HARM

> S: It just hit me . . . we have lost a lot of people this year already.
>
> T: Yeah. I was watching a series about gangs, and I realized that there are only a few groups of people who lose people as frequently as we do. Soldiers, gangs, and us. It struck me that those groups lose people to violence so often, where that is not the case with us . . . but yet the life we lived was often like war. Maybe more chaotic. I feel like all the loss makes me insensitive to it, and then it happens again and I am devastated.

I next introduce one of the more macabre issues involved in my sampling. As a social scientist, I am interested in understanding the holistic experience of living within the community of Bourbon Street women. I am fortunate to have some interviews with women whom I have known and maintained contact with for more than two decades. One topic of conversation that is startlingly common is death.

Of the 71 women who made up my initial sample, at least 4[1] are reported within the group to have died of homicide. While it might be tempting to dismiss this (almost 6 percent) homicide rate as the downfall of a small sample, simulations have shown it to be significantly higher (p less than .001) than the concurrent U.S. rate of 6.3 per 100,000. At least 2 others have died of drug related deaths. Another was the victim of a domestic violence case so brutal it made national news, years later leaves her unable to function and dependent on disability checks for her survival. My research indicates that women in my B-drinking sample are dying violent deaths at a higher rate than the general public.

Women face violence and coercion in diverse ways in American life. Women in different relations to sex work (here comparing sex workers and customers' wives) "encounter varying shades, forms, and tonalities of power and violence against them . . . they stand in different relation to the law, to safety, to convention" (Price 2012, 65). This is supported by the rate of violent death for women in my sample, which is far higher than the homicide victimization rates reported by the U.S. Department of Justice. According to Alexia Cooper and Erica Smith (2011, 2), reporting on homicide patterns and trends in the United States from 1980 until 2008, the number of homicides reached an all-time high of 9.8 per 100,000 people in 1991. By 2010, the rate had declined to 4.8 homicides per 100,000. My data suggests that B-drinkers, in comparison, are victims of violent homicide at a rate of more than 5,500 homicides per 100,000 B-drinkers. Although the National Crime Victimization Survey of the Bureau of Justice Statistics in 2001 did not identify the exotic-dance workplace as at risk for violence (Hanna 2012, 213), none of the women in my sample were killed in their workplace.

It is disheartening to me that death has become such a commonplace theme among this group of women in their twenties to fifties. The topic is ordinary enough to generate frank discussion, both inside and outside the context of interviews. Conversations about past events lead to discussions about where other women are today, and with alarming frequency I hear stories about the deaths of very young women.

DOMESTIC VIOLENCE

It is difficult to gather reliable data on whether people who engage in sex work are more likely to be victims of domestic violence, both sexual and nonsexual, than those who do not. Reporting of such violence is low, making qualitative, ethnographic observations necessary to understanding the experience of discriminatory violence against sex workers like the B-drinkers of the French Quarter. The vulnerability to domestic violence I found in my sample adds to the literature on North American strip-club employees and violence perpetrated against them by those with whom they are involved outside the workplace. My research suggests that the rate of reporting is lower among women employed as B-drinkers than other populations because of the legality of their professions.

It has taken many years for my informants to reveal to me the pervasiveness of domestic violence among B-girls employed during the 1990s in New Orleans. This is, in part, because American women tend to hide their own victimhood in cases of domestic violence broadly; during 1993, I heard dressing-room talk about domestic violence no more or less than among graduate students, or adjunct faculty, or any other group of which I was a part. However, a sample of 71 of any of those other groups does not generate such a startling mortality statistic. This is probably due to a combination of job stigma, socioeconomic status, and the people that associate with the women both inside and outside the clubs.

In the United States, people tend to engage romantically with those from their own social class (Whyte 1990; Gilbert 2010, 110). In the case of the striptease dancer and B-drinker, it is often the case that working-class boyfriends and husbands earn substantially lower incomes than their partner, inverting gendered expectations about American households. This may contribute to increased conflict within relationships. Valerio Simoni (2015) writes that, in Cuba, men who engage in sexual relationships with tourists face the dilemma of expressing culturally different configurations of masculinity within and outside the tourist realm. Similarly, B-girls in New Orleans must negotiate between multiple femininities, as their intimate encounters with local partners differ from their intimate encounters at work.

Successful B-drinkers must meet the requirements of the professional businessman's desires and perform an upper-middle-class man's fantasy. This includes attentiveness, sexual availability, and specific styles of dress and makeup. These differed greatly from the femininities expected within their own social circles, particularly for the youngest women. It was often the case that women leaving the club in their street clothes were unrecognizable to customers, having removed their long wigs, which revealed unconventional hair colors or asymmetrical short haircuts; wiped off the foundation covering their tattoos; and reinserted various piercings. Beyond appearances, shifts in dialect and body language, changes in affect from seductress to street-tough, and changes from cooing dependent to financial head of household led to a "criss-crossing traffic of different models and vectors" of femininity and identity (Simoni 2015, 407).

The stigma of B-drinking and working in a strip club adds to the vulnerability of such women to domestic violence. Partners of B-drinkers are not immune to the influence of the American whore stigma, and their expectations of feminine sexual exclusivity rather than availability led to arguments and sometimes physical violence. Some women reported having difficulty with boyfriends who both wanted them to quit, and also depended upon their high incomes for support. The community of men willing to engage in long-term relationships with women employed on Bourbon Street was, I believe, one of the problems faced by such women.

> It's marginalizing. And so it limits you in your relationships. God . . . I can't even describe how fucked up he [her live-in boyfriend] was. But, um, yeah I supported him our whole relationship. But I just started dancing when I first met him, and you know his mom was a stripper, prostitute, biker gang . . . so he had. . . . That was family life. So, um, you know you did have girls who did that. And then you had the sort of . . . the men that would sort of like, the little bottom feeder men, you know, who, like, would sleep with all the girls at the bar, and like claim that "I love dating strippers, because they support me, and I'm like a Rock Star,"—Jade, interview, 2012

Aside from being deadbeats, some of the men who dated dancers lived socially marginalized lives outside of their relationships. During interviews done in 2013 and 2014 especially, the nervous laughter of some of my friends seemed deeply misplaced when they spoke about situations of violence that they had experienced themselves or heard about through the group's various networks. One woman, in an interview done in the summer of 2014, relayed to me a story concerning the eviction of a past boyfriend from an apartment in 2001. When the landlord found the empty apartment, the refrigerator was unplugged and "filled with animal heads," apparently left as a commentary on his feeling about their landlord/tenant relations.

Although I do not suggest that all sex workers' partners are dangerous or criminals, some of them clearly were. What was striking to me in my most recent interviews was the desensitization of my informants when talking about horrific acts (to me) that occurred in the past. This is an example of a story told to me in 2013, concerning the boyfriend of a girl who I did not know but who was also a B-drinker.

> Well, she lived around the corner and she had a son, that I babysat until I got lice from him, and I was like I can't it's a liability and I have the key to use their washer and dryer and then . . . I hadn't seen them in a bit and I just . . . I just like, I just got a bad vibe when I went to return the key, and I'm like "I don't need it anymore I do my laundry at my boyfriend's (A: right right) "thank you," and he didn't let me in, her boyfriend.
>
> I found out later on when they got evicted, um, they had moved the body that was under the house [nervous giggling] and dumped it in the swamps and they found it then. But it was a roommate, who worked at [fine dining restaurant], and they had a fight, and . . . killed him in front of the little boy, and then just left the body underneath the house.
>
> And then, you know . . . and it was like smelling, and when I went over there to return the key, and they weren't letting me in it was there and Kittens said she'd been IN the house and they were like "I think a rat died." . . . It was like [nervous laughter] . . . it was just so . . . [laughter]

COMMUNITY, LOSS, SAVANNAH

I call this informant Savannah. She was in her early twenties, a normal age for a dancer. She was, however, on the borderline of developmental disability by our cultural standards. Although admittedly I never administered an I.Q. test to Savannah, I have worked in both classrooms and group homes with children and adults receiving special living assistance before becoming an anthropologist.

Savannah exhibited many of the mannerisms I had witnessed in other women struggling with age-appropriate behaviors. She also had poor math skills and consequently had difficulty making change, even when dealing in multiples of five. She had weak social skills and often did not understand social cues to other people's points of view. B-drinking as a type of work quickly teaches a person to look for social cues suggesting that customers are or are not interested in her, and Savannah did not perform well in this area even after months of observation. She also had difficulty managing her body. Although her hygiene was good, she could not care for her hair or clothes and had difficulty keeping track of her belongings. Fortunately, Savannah's clothes, such as the blue nylon, puffy winter jacket she wore when cold between dance sets, were unappealing to the other dancers. This minimized her personal losses to theft in the insecure dressing room, where bags were constantly riffled by employees and employers alike.

Savannah was an attractive woman. Bonding groups of female employees in the dressing room on nights without customers would often (perhaps weekly) curl Savannah's bleached-blonde hair, lavishly cosmetic her facial features, and dress her up in an amalgam of their less-favored costumes. Such scenes of leisurely play were common in the dressing room. Sexual tension between the girls, who are much more likely to go home with each other than with any customer, are sometimes played out in the dressing room, and sometimes as campy, hyperbolic performances during later floor shows for tips.

Savannah was a dancer for at least two years and was a fixture in one of the clubs where I made observations. She was always the object of the above-mentioned makeovers, never the subject. Although her behavior was sometimes annoying to customers, usually the other women helped to steer her into appropriate situations. Except for a few loud arguments, usually when the angry dancer was drunk, the girls tended to recognize and assist when Savannah needed help.

Savannah was functional as a stripper and could earn enough money to support herself well. There are very few jobs which pay unskilled female laborers as much per night in the region, and young women drive to New Orleans from the surrounding states in surprising numbers. Often, Savannah was the lowest-paid girl at the end of the night. Still, every girl has her high-income nights, and when Savannah had hers they were quite good. There are various patterns of income that dancers learn to expect. Based on my estimates and our conversations, Savannah earned perhaps $80 per night, with a $300 night perhaps once every three weeks. Such good nights were usually generated from only one or two customers.

I knew Savannah sometimes met customers outside of the club for "parties" in their hotel rooms. Usually, a "party" goes on for most of the night, with the customer providing a room, drinks, cigarettes, and between $300 and $1,000. The girls perform private dances, alone or in pairs, with ongoing negotiations often culminating in actual sex. Several women I knew performed in pairs or small groups, spending most of the party having sex with each other (a better strategy in the contemporary environment of HIV). These performances rely strongly on the tradition of "camp," here the self-conscious parody of sexualities, exaggerated performance that serve as a form of rebellion against the codification of heterosexual norms. "Camp sees everything in quotation marks. It's not a lamp, but a "lamp;" not a woman, but a "woman." (Sontag 1966, 280; see also Erik Anderson, 2010). These exaggerated acts of "sex" did not preclude actual affectionate relationships between these women, who often pair up with friends to best protect themselves from dangerous individuals who prey on sex workers unable to defend themselves because of the illegitimacy of their work. The strong community ties are likely in part a response to this cooperative protection.

Savannah died in an apartment, where she and another dancer were attacked. I was told that Savannah was first struck from behind with a blunt object, knocking her unconscious. She was later found in a trash bag, in a distant neighborhood. I will avoid describing the specific disfigurement of Savannah's body in order to protect her anonymity. The police unofficially questioned me in the bar one night about Savannah. The detective came in and showed me a series of photos of women, asking me if I knew any of them. At a certain point, I blurted out, "Hey! That's Savannah!" That was when I learned that Savannah really was dead, and it wasn't just a rumor.

This event was the focus of dressing-room narratives for weeks. The stories always reported that Savannah left The Club with the other girl involved in the attack (whom I did not know but will call X). Several girls reported having witnessed this. Girls thought they had left with a couple that X knew but Savannah did not. They were planning to have sex, and several girls confirmed to me that Savannah, at least, was going to be paid for her participation in the party. No one knew anything else. No one could tell me where X was, although she was believed to have survived the attack. No one knew the couple, although there was general agreement that the woman had been a dancer somewhere on Iberville Street, in one of the low-end clubs. She was now gone.

This is the danger of sex work. Why was no one watching out for Savannah? It is a question that goes beyond our culture's explicit statement about illegal sex work. There is a tacit understanding that such women are disposable, and this institutionalized belief puts women's lives in danger.

POLICE VIOLENCE

The understanding that sex workers can be abused is firmly entrenched in the practices of street police. The police officers in the French Quarter have the power to arrest any of these women at any time, on charges varying from public drunkenness to lewd dancing. B-drinking is such a charge, a crime that makes all women in the clubs vulnerable. This is a space where victimless crimes occur, like lewd dancing, with such vague definitions that they require a lot of subjective decision making on the part of the arresting officer. During my years observing in various strip clubs in the French Quarter, these are some stories that I was told or witnessed:

> Jenny was "working on the door" (which means that she was standing in the doorway of the bar trying to start a conversation with a passerby to advertise a private dance. Although the dance is legal, soliciting one from the door is not, putting Jenny in a vulnerable position). A man comes up to Jenny and express-

es interest in going in the back room with her. She is trying to negotiate a price to go in the back room for half an hour (where there is a sofa), drink a bottle of champagne, and give him unlimited private dances with no contact. The man is very drunk and wants Jenny to leave the club with him. He flashes a wallet full of a badge and money, and Jenny again suggests he come in and spend his money in the VIP room. Realizing that she will not leave with him, he begins berating her as a "whore" as well as other pejoratives revolving around her open sexuality as a sex worker. Jenny returns his aggressive attack, calling him pathetic and unattractive. Grabbing her by the shoulders, he pushes her up against the door and growls into her face that he's going to see her dead in the bayou. At this point, the bouncer breaks up the fight, convincing the man that the girl is just drunk (when she is not) and joking with him before sending him stumbling down the street. The bouncer then sent the girl home for two weeks, suggesting that the individual (whom he had recognized) will have forgotten her by then.

Mercedes was off work on a Tuesday night, but was sitting at the bar drinking for free. The bartender was a friend, and she was upset about a relationship with a former boyfriend. Joe, a security guard and retired police officer who regularly visits the bar, walks in. He is in uniform and has been doing private security work at a nearby hotel. Mercedes knows Joe, and they have often passed friendly words, and he stands next to the bar and listens to her complain to the bartender for a moment. Hearing that Mercedes' relationship with her boyfriend was over, Joe walks up behind Mercedes, sliding his hands around her body and into her bra through the sleeves of her tank top. He simultaneously begins talking in her ear and holding her breasts. She jumps up and runs into the dressing room, clearly upset. He laughs, and the bartender gives him a soft drink. He laments that he can't drink because he is working.

A. J. is a French Quarter cop. He came to The Club about once a week, and often cornered particular girls, purchasing them one small cocktail to get them to sit with him. Sadie was a waitress who was not interested in B-drinking or other forms of sex work. She would not sit with customers, she refused to dance. She was pretty and a quick waitress. A. J. developed an interest in her, and over the course of a month came in with increasing frequency. He began buying her drinks, making small remarks about the legality of her job and insisting that she sit with him. After the first week he began pulling her into his lap, often when drunk. Dancers who sympathized with Sadie tried to intervene by inserting themselves, but A. J.'s attention was not budged. After about one month, Jenny began working at another club, hoping to hide from A.J.'s attention. Eventually, she left Bourbon Street altogether.

These narratives and the way they were described to me during interviews revealed that women working in the New Orleans sex industry were open to aggressive and sometimes violent displays of power by local police. The attitudes of the police are indicative of the population in general. Young women whose labor supports the internationally known Bourbon Street environment of bawdy entertainment are closely monitored and work with tight restrictions on their behavior, but those restrictions have very little to do with morality or the law.

The hostility of this work environment seems to have long-term effects. In 2015, I administered a survey concerning the overall health of women in my sample population. This Wellness Survey was aimed to specifically look at non-communicable disease, rather than contributing to fears of sex workers as vectors of disease. The survey included responses from 10 women who had been B-girls for 3 or more years, with 9 aged 36–45 and one older than 46. Of those who completed the survey, their current occupations ranged widely, including health care providers, teachers, restaurant managers, and technical support. Four identified primarily as homemakers, but two of those worked part-time as a substitute teacher and yoga instructor. All had completed at least high school, four had 2-year degrees, two had 4-year degrees, and two had terminal professional degrees. The breadth of outcomes seemed to defy patterns—except for the category of anxiety-related illnesses. Nine of ten reported suffering from anxiety, six reported PTSD, and four reported suicide attempts. This is despite all 10 reporting currently living in a situation where they felt safe in their home, with their partner, and with all other family members.

> I carry that with me today, like I still feel like uncomfortable, you know, being hit on in bars, like I very much just like wanna . . . you know look nice, and have people admire me from afar, but I DON'T want you to talk to me. I don't want you to acknowledge that I am this woman who can use my sexuality to get what I want, because I am NOT that woman, I haven't used it to get what I want. I use my brain. Like, that's why I'm in [current job] . . . it's very . . . clear? You know? That [being a B-girl] is not me. I think in the long term, it's affected who I am very much so, but in ways I've tried to distance myself from being that person, you know? But I still maintain a sort of need to be, like, on the fringes of society, because I can't function normal . . .

Among my interview participants is another woman (like me) who has gone on to become a social scientist. At the end of our first formal interview, she talked briefly about attempts to return to the clubs on Bourbon Street to engage in qualitative research. We both agreed that it was uncanny how difficult it was to approach those same clubs now, despite overwhelming feelings of nostalgia for past times there. We had both had similar experiences in trying to do contemporary research within the clubs. Here is our conversation:

> P: I could have done this project, but it just became . . . I also had a problem . . . I did get a job as a waitress, but I couldn't go, I couldn't do it. I just couldn't bring myself to do it . . .

A: Like you couldn't go to work, once you had the job . . .

P: Yeah. And then I was like, oh, the only way I can do this is by dancing again, and I considered it many times. And I got drunk just to do it, and I went downtown, but I couldn't do it . . . .

PHOTOGRAPHS AND AGENCY

One form of control that I have already discussed is the visual monitoring of behavior. In two of three clubs that I visited, I found that the manager's office was equipped with hidden cameras that the employees were unaware of. Young women in the peep booths, on the stage, and urinating in the bathroom could all be watched from the comfort of the manager's office chair. Management can thus monitor the observable behavior, compared to reported behavior, of B-drinkers. These cameras, which some women believed produced films that were marketed without the model's consent or even knowledge, are just one way that the labor of young women is stolen by modern clubs, serving as a postmodern sort of techno-pimping.

Other participants in the industry have complained about the misappropriation of imagery through uninvited photography and filming by customers of clubs (see, for example, the documentary film *Live Nude Girls, UNITE!* by Query et al., 2001). As elsewhere, young women working as dancers in New Orleans struggle to maintain control over their images. A common confrontation between customer and dancer is over photographs being taken without the dancer's consent. In a club walled in mirrors, a camera flash is obvious and can purposefully or accidentally capture the images of many people in various corners of the property.[2] Part of the waitress's job is to temporarily confiscate the camera to give the dancer the opportunity to negotiate a price for the customer to be allowed to keep his film. Her image is her income, and photographs are commodities in the club. Still, one is generally allowed to take photos in public places, and I have witnessed many arguments with litigious American men about the visibility of the "no photos" sign, the value of other photos on the roll of film, etc. The argument ultimately is about whether the woman should have power over her image, or whether as a strip-club employee and thus obviously a woman of loose morals she has lost that right.

In *Live Nude Girls, UNITE*! one of the dancers noted that in San Francisco, dancers who complained about their photos being taken against their will were told that if they felt that way, they shouldn't work there. The dancer, in the process of early negotiations for the first unionized exotic dance club in the United States, replied "No. I do work here, and I do feel that way." In New Orleans' strip clubs, women who get angry about being photographed by customers are told by management, wrongly, that this is an unavoidable job hazard. This tacit understanding in American culture that women who work in the sex industry should expect to lose control of their ability to say yes or no to any particular act (here, the woman is willing to do live shows, but unwilling to work in pornography, or the graphic depiction of similar acts) illustrates the notion that women working in this industry are vulnerable to loss of basic civil rights. I say "vulnerable," because without the structural support of being classified as a club employee, and without the decriminalization of prostitution, her work is not protected by the same rules that apply to other workers in blue- and pink-collar jobs in our society.

It would be wrong to oversimplify the relationship between photography and sex work and imagine there is only power and control on behalf of men at work. Several women that I worked with were models for the popular *Girls Gone Wild* video series while they were filming in New Orleans, and proudly wore their tee-shirts like trophies. Although they were not paid for this work, the high-profile advertising of the videos could further legitimize a dancer's sex appeal. During the months that universities have spring break, for example, the combination of the pink shirt, tanned skin, and perhaps underpants with a tropical print were sufficient to guarantee enough face appeal to maximize the income from that season. Here, although certainly the video production company benefits the most, the women involved are not victims but subjects who maximize the use of their image in various ways as a form of capital in the industry. The lines between sex work and pornography modeling are thin.

I've suggested two things in this chapter. The first is that my observations in New Orleans suggest that women working in the sex industry are, although not powerless, working in dangerous situations over which they have less control than women as a general category in the United States. The history of sex workers' rights suggests that at times when women more broadly have increased power in society, women in the sex industry have less (Alexander 1998, 185), an unfortunate negative correlation. Although women in the United States are increasingly empowered as we have increased understandings of gender, we must proceed with caution. Here, a group of women (who are overrepresented in populations already in positions of vulnerability, such as runaway abused minors, single mothers, and poor women) are actively disempowered both in and, perhaps because, of their work environment. They must constantly protect themselves against discrimination and violence.

NOTES

1. I say "at least" because of another case that was brought to my attention during interviews. A woman officially reported to have committed suicide decades before was believed by

my informant to have been a victim of death by erotic asphyxiation after which her partner fled the scene. She reported that investigation of the case was not pursued further by the woman's parents, who arranged for her funeral. The combination of stigma and the grief of family members contributes to the difficulty of investigating such cases, again providing barriers to our understanding of the numbers of women affected by such violence.

2. Undoubtedly this problem is even worse for young women working today. With smartphone technology and the ability to instantly store an image in the cloud, the 1990s B-girls' most extreme strategy of destroying a tourist's camera would not necessarily protect her today.

Conclusion

By the 1990s, "B-girldom" had developed as a discourse mobilized by those in power to stigmatize and control women who work at the social margins of the tourism industry. Their work as dancers and the more intimate work of B-drinking is stigmatized because it is a manifestation of the intersection of (performed) sexuality and the tourism economy. Though few women use "B-girl" as an identity category for themselves, the term is used widely to label and discuss any women who practice this strategy as part of their work. The creation of the legal category of "B-girls" reflects a long history of defining and controlling female service-industry workers who violate perceived gender roles as they "should" play out according to American heteronormative expectations about gifted drinks, alcohol consumption, and sexual availability.

Solidified in both media and political discourse, notions of B-girldom as an identity have evolved in the context of the New Orleans' touristscape. Although women in the 1990s did not self-identify as B-girls, the *strategy* of B-drinking remained a common strategy in strip clubs frequented by visitors to the French Quarter. Individuals whose jobs require B-drinking as a component run the risk of such identification, consequent arrest, and a range of other types of violence. Geographically distant but theoretically similar, Holly Wardlow (2006) dissects the term *pasinja meri* in the cultural context of New Guinea, showing that the referent has developed to control increasingly uncontrollable women and serve goals of gender hegemony specific to the New Guinea context. Likewise, the construction of the B-girl serves to minimize the hard economic power held by contract workers who perform on Bourbon Street.

Stakeholders in the tourism economy have, for decades, struggled to control the agency of such women in order to maintain overall control of Bourbon Street, a focal point in the touristscape. The role of the B-girl, like all stock characters, is adjusted to her appropriate moral positioning in storytelling at specific historic moments. Notions of B-girldom as they evolved can be traced through a look at the discursive B-girl in newspapers, movies, plays, and the legislature over time. Stories about B-drinking reflect local shifts in attitudes toward the sex industry more broadly in the city from 1941 until 2012.

B-drinking has been a named feature since at least the 1940s, although the strategy clearly predates legislative attempts to control such workers. The media has functioned alongside legislation in the process of creating B-girldom. Sources like the local newspaper *The Times-Picayune* reveal the process of "making up" (Hacking 1986) the B-girl. This discursive trick of society serves to disempower and disenfranchise a known and necessary group of young female stakeholders in the French Quarter as a heritage tourism site.

This discourse of B-girldom was created in a top-down manner by powerful stakeholders in the city's tourism development; New Orleans' elites have long recognized the market for intimate encounters between visitors and French Quarter inhabitants, and sought to control and benefit from such interactions. Through law and media, the ways that sex workers of various kinds have been permitted to engage with visitors to the French Quarter have been molded. For the city's inhabitants, such engagements have become naturalized because of their frequency in the public discourse. Authorities outside of the French Quarter community have sought to control how sex workers on Bourbon Street are understood and accepted by the local people of New Orleans. Historical documents, such as the archives of *The Times-Picayune*, reveal that B-drinking is a normative but contentious aspect of the tourist economy of Bourbon Street as it exists today and has historically been constructed.

Denise Brennan (2004) and Erica Williams (2013), in discussing the Dominican Republic and Bahia respectively, illustrate examples where the constructed hyper-sexualization of a touristed site's inhabitants are related to issues of race that stem from the unique historical, colonial, and social situations. Similarly, I suggest it is folk notions of a history of vice that led to the formation of New Orleans as a "site of desire" (Manderson and Jolly 1997), "a place that is eroticized and imagined as a site for romance, sexual adventure, and license" (Williams 2013, 5). Although race is involved in New Orleans, it is only one component of the exotic that has molded New Orleans' place in the global sexscape. Not every tourist to the city visits a strip club, but the existence of such places remains a critical component of the atmosphere of the French Quarter in the imagination of the tourist. The role of New Orleans as a site of desire is part of what intrigues visitors, and the city's social sexual history (Frohlick 2007) predates modern tourism and the boom of modern strip clubs, which are perhaps themselves secondary effects of expectations that evolved with the port. The B-drinker, B-girl, or Bee girl is a part of the tourism imaginary, an imagined person that has "become culturally salient in tourism" (Leite 2016, 264), though imagined by the courts, writers, and people of New Orleans before being imagined by the tourist.

Arjun Appadurai's notions of the imagined place are important to understanding the sexual economy of the French Quarter and B-drinking. Tourists understand New Orleans' French Quarter as it has been represented in the mediascape, in written narratives, and in film. Intimate interactions with visitors are carefully controlled by the New Orleanian ideoscape as it is constructed within the legislature and the courts. Together, these two separate discursive spaces contribute to the construction of the French Quarter as a particular type of touristic space, one where vice and revelry are expected. New Orleans' weak position in the U.S. economy additionally contributes to its appeal as a sex tourism destination where cheap labor, including sex work of many kinds, can be found.

The tourist as consumer is not the only voice that controls what will be supplied at any tourist destination. Visitors who serve as "money pumps" (Donlon and Agrusa 2003, 125), introducing new money (rather than recirculating it) into the city's economy via the touristscape are not the only influences controlling the nightclubs, walking tour companies, and busking musicians that make up the cohesive community of the French Quarter. The inhabitants of the city broadly, and particularly the economic and political elite, are important coconspirators in B-drinking and French Quarter antics. In order to fulfill the demand, the community must agree to supply.

Discursively, the B-girl has been constructed, moralized, and controlled by narratives in the local newspaper as well as fiction written by local authors. A second discourse about B-girls exists in the Louisiana legislature, city ordinances, and appellate documents. Together, these local constructions of B-girldom work to influence the average New Orleanian's understandings of the relationship between women working in the service industry and visitors of all sorts.

The French Quarter, like any heritage tourism site, is part reality and part construction. The created experience of touring the French Quarter both controls and is created daily by a wide variety of tourism stakeholders, reflecting both tourist demands and local moralities. This dialectical construction of carefully selected past moments of preservation create the local touristscape and its perceived available opportunities. New Orleans, though a still expanding touristscape, remains one whose backbone has always been the expectation of vice broadly, and sexual permissiveness particularly. This is a part of New Orleans which is inseparable from other aspects of its marketable past. Women working in nightclubs as striptease performers and B-drinking companions are crucial and significant actors in this touristscape.

Understanding the contemporary sexscape of the United States as it intersects with and is a component of the touristscape of the city of New Orleans is enriched by a historic study of practices particular to the city. Peirce Lewis (1976, 11) wryly suggested that "there is scarcely any point for a scholar to study New Orleans, particularly if the scholar is predisposed to study cities, in search of general urban theories." New Orleans occupies a very specific niche as a port city, a past colony of France and Spain, a Catholic city, and a city in a precarious geological niche in constant threat of floods. It is for exactly this reason that New Orleans can add to our understanding of the development of the American sex industry. The case of the quasi-acceptable B-girl underlines the importance of understanding the broader history and political economy of a region in order to make sense of how the sex industry intersects with tourism in that place. The history of the city and the response of its residents over time have created this unique approach to engaging in sex work. It is through such seemingly different examples as the B-girl that the constructed nature of tourism interactions are revealed.

The precarious positioning of service workers who are required to B-drink reflect city officials' attempt to control an otherwise independently organized work force. In response to such restrictions, women whose work included B-drinking in the 1990s strategized as a group to maximize their income as individuals. This need to work cooperatively levels the playing field among individual contract workers, creating a sense of comradery among those who might otherwise compete against one another. Restrictions meant to control B-drinking, in the end, result in the creation of communities of women who are united by their shared stigma. Working together, they can and do use their collective power to control the pricing and availability of the forms of sexual encounters available to customers on Bourbon Street.

Longitudinal social research on the lives of sex workers of any sort is rare. In this case, because of the intense community ties formed within the group in response to their ambiguous legality as strippers/B-drinkers, cliques of women who worked together in the 1990s continue to function as unified groups as long as twenty-five years after leaving Bourbon Street. Women who worked together in the 1990s continue today to extend the use of fictive kinship terms (such as "Mama"), assist one another in times of financial crisis, and provide emotional support to one another. This study suggests that further research into the long-term bonds created by stigmatized workers is warranted.

Important to understanding the community's need for long-term group affiliation is evidence I discuss in my final chapter. B-drinking is a solid advancement strategy for otherwise unskilled workers in the tourism economy, some of whom use this work as an opportunity to obtain capital to start businesses or attend school. However, longitudinal research reveals that B-drinkers inhabit a position of structural violence. There are significant, though difficult to measure, differences in rates of violent death among B-drinkers and the general populace of the United States. The need to discuss the loss of companions to homicide, domestic violence, and drug overdose are met by the group; conversations that are normalized within group communication are extraordinary within the broader social contexts of the United States. I believe the use of social networking by previous B-drinkers fulfills personal needs to process such violence as a member of the group rather than alone. Years after leaving Bourbon Street, the members of the community rely on one another for opportunities to reflect on their experiences with those who have shared similar experiences and who view gender roles and gender violence through a lens that is relative to their own.

I conclude this book with some suggestions on how my study can be applied to the contemporary situation in the French Quarter. I do not support the suppression of businesses like The Den, which I view as providing opportunities rather than contributing to problems within the New Orleans touristscape. Establishments like The Den do not create secondary effects (Renton 1986) in the surrounding communities, but instead buttress the tourism industry broadly. It is the stigma of exotic dance, B-drinking, and other forms of sex work that leaves such women as workers vulnerable to danger. I would urge the community of New Orleans to provide additional services, such as drop-in houses for youth and shelters for victims of domestic violence, to address the specific problems faced by the women who work to provide the intimate services desired by the tourists that support the economy of the city. Ultimately, it is not B-drinking that harms such women. It is the stigma and discrimination against such workers that leads them into relationships and situations where they face dangers.

Bibliography

1943a. "Ban on 'Bee-Drinking' Voted Here; Provost Officer Sees Action Taken." *Times-Picayune*, January 23, 1943, 1.
1943. "Police to Enforce Anti-Bee-Drinking Order, Says Superintendent Reyer." *Times-Picayune*. January 28, 1943.
1943b. "Bee-Drinks Out; Girls Switch from Slinky Gowns to Waitress Aprons." *Times-Picayune*, February 2, 1943, 1.
1943c. "Holmes Condemns Bee-Girl Practice." *Times-Picayune*, February 11, 1943, 23.
1943d. "This is war, Private learns upon furlough trip home, Rigors of civilian life too much for soldier." *Times-Picayune*, April 14, 1943.
1944. "Plan to control vice formulated, City leaders agree to tighten up rules." *Times-Picayune*, March 29, 1944.
1945. "All Under 18 Prohibited in Barrooms, Says Reyer: Reports Age Limit Decided After Check of Laws." *Times-Picayune*, February 7, 1945, 1.
1946. "Club is Closed Operators Held, Arrests Follow Charges of Bee Drinking." *Times-Picayune*. July 16, 1946.
1946a. "Manager of Club is Judged Guilty: Allegedly Hired Minor as 'B' Drinker." *Times-Picayune*, October 24, 1946, 4.
1946b. "Views on Sundry Topics: Could This Be Sarcastic?" *Times-Picayune*, May 13, 1946, 8.
1947. "Police Reports: Eight Arrested on 'B-Drinking' Charge." *Times-Picayune*, February 25, 1947, 21.
1948. "Wife 'Lured' Robbery Victim, He Charges." *Times-Picayune*, June 24, 1948, 40.
1949. "Carnival King Reaches Mouth of River; Will Land Tuesday to Lead Festivities: Bar Pilot." *Times-Picayune*, February 27, 1949, 2.
1950a. "'B-Girl' Vanishes; Life Threatened: Woman Believed Involved in Extortion Case." *Times-Picayune*, April 6, 1950, 1, 12.
1950b. "French Quarter Committee Votes Faith in Gulotta." *Times-Picayune*, April 13, 1950, 2.
1954. "Liquor Licenses May Be Revoked." *Times-Picayune*. December 16, 1954.
1955. "Night Club Job Told by Girl, 13." *Times-Picayune*. January 5, 1955.
1955. "11 Persons Held After Bar Raids." *Times-Picayune*. July 17, 1955.
1955. "Five are Booked in 'B-Drink' Case." *Times-Picayune*. October 14, 1955
1950c. "Mayor to Name Vice Inquisitors, Ask B-girl Ban." *Times-Picayune*, March 30, 1950, 2.
1981. "Movies on TV: Briefly—'The Man with the Golden Arm.'" *Times-Picayune*, December 6, 1981, 7.

1985a. "Legislative Briefs: Local, a Bill Passed Unanimously." *Times-Picayune*, May 16, 1985, A30.
1985b. "Legislative Briefs: Men Who Are Paid." *Times-Picayune*, June 29, 1985, A30.
1989. "6 Arrested in Raid at Harvey Bar." *Times-Picayune*, June 20, 1989, B3.
1992. "Women's Charges Wrong." *Times-Picayune*, October 14, 1992, B8.
1993. "Police Reports: Terrytown." *Times-Picayune*, June 22, 1993, B3.
2011. "1944: Higgins industuries in New Orleans contributes to war effort." *Times-Picayune*, November 17, 2011. http://www.nola.com/175years/index.ssf/2011/11/1994_higgins_industries_in_new.html.
Alexander, Priscilla. 1998. "Prostitution: Still a Difficult Issue for Feminists." In *Sex Work: Writings by Women in the Sex Industry*, 2nd ed. Edited by Frédérique Delacoste and Priscilla Alexander. 184–230. San Francisco, CA: Cleis Press
Anderson, Erik. 1979. "Lay Winning Vice Battle: Bossier City's 'Strip' Fading Away." *Times-Picayune*, August 13, 1979, Sec 1, 1, 12.
Anderson, E. 2010. "Sailing the Seas of Cheese." *Contemporary Aesthetics* 8. http://hdl.handle.net/2027/spo.7523862.0008.019.
Appadurai, Arjun. 1990. "Disjuncture and Difference in the Global Cultural Economy." *Public Culture* 2 (2):1–24.
Appadurai, Arjun. 1996. *Modernity at large: Cultural dimensions of globalization.* Minneapolis: University of Minnesota Press.
Ashforth, B. E., Kreiner, G. E., Clark, M. A., Fugate, M. 2017. "Congruence work in stigmatized occupations: A managerial lens on employee fit with dirty work." *Journal of Organizational Behavior* 38(8): 1260–1279.
Ball, M. 1986. "A shocking display of mystery-writing talent." *Times-Picayune*, April 20, 1986, C10.
Berg, Bruce L. 1989. "A Dramaturgical Look at Interviewing " In *Qualitative Research Methods for the Social Sciences* Boston: Allyn and Bacon.
Bourdieu, Pierre. 1986. "The Forms of Capital." In *Handbook of Theory and Research for the Sociology of Education*. Edited by J. Richardson, 241–258. New York: Greenwood.
Bourdieu, Pierre and Loïc J. Wacquant. 1992. *An Invitation to Reflexive Sociology*. Chicago: The University of Chicago Press.
Bourgois, Phillippe. 1996. *In Search of Respect: Selling Crack in El Barrio*. Cambridge, UK: Cambridge University Press.
Brennan, Denise. 2004. *What's Love Got to Do With It?: Transnational Desires and Sex Tourism in the Dominican Republic*. Durham, NC: Duke University Press.
Brinkley, D. 2000. "The man who won the war for us: The neglected epic of Andrew Jackson Higgins." *American Heritage* 51 (3):49–50.
Bruner, Edward M. and Barbara Kirshenblatt-Gimblett. 1994. "Maasai on the Lawn: Tourist Realism in East Africa." *Cultural Anthropology* 9(4), 435–470.
Bus Stop. 1956. Joshua Logan, director. Twentieth Century Fox.
Butler, J. 1871. *Letter to My Countrywomen.*
Campanella, Richard. 2008. *Bienvillle's Dilemma: A Historical Geography of New Orleans.* Lafayette: University of Louisiana at Lafayette.
Campanella, Richard. 2014. *Bourbon Street: A History*. Baton Rouge: Louisiana State University Press.
Centre, The Social Issues Research, ed. 1998. *Social and Cultural Aspects of Drinking*: The Social Issues Research Centre.
Chu, Kay H., Melissa A. Baker, and Suzanne K. Murrmann. 2012. "When We Are Onstage, We Smile: The Effects of Emotional Labor on Employee Work Outcomes." *International Journal of Hospitality Management* 31(3), 906–915.
City of New Orleans Mayor's Office Press Releases. 2014. New Orleans Achieves 9.28 Million Visitors in 2013. http://www.nola.gov/mayor/press-releases/2014/20140422-tourism-numbers/. Accessed February 10, 2015.
Clinton, E. M. 1949. "Bone to Return to Mississippi." *Times-Picayune*. February 17, 1949
Coleman, Leo, ed. 2011. Food: Ethnographic Encounters. New York, NY: Bloomsbury Academic.

Cooper, Alexia and Erica L. Smith. 2011. *Homicide Trends in the United States, 1980–2008*. U.S. Department of Justice, Office of Justice Programs, Bureau of Justice Statistics.
Counihan, Carole M. 1999. *The Anthropology of Food and Body: Gender, Meaning and Power*. New York, NY: Routledge.
Counihan, Carole M. 2004. *Around the Tuscan Table: Food, Family, and Gender in Twentieth Century Florence*. New York, NY: Routledge.
Cuthbert, D. 1998. "Allouette—The Tassel Queen of Bourban Street is Still Jaunty After—All These Years." *Times Picayune*, January 31, 1998, E1.
Cuthbert, D. 2000. "Frankie Speaking—From Bourbon Street to Lenny Bruce and 'Dracula's Dog,' Frankie Ray Has Seen It All." *Times Picayune*, January 15, 2000, E1.
Darby, Joe. 1996. "Metairie Veteran Stormed Beaches." *Times-Picayune*, November 11, 1996, B1.
Deegan, Mary Jo. 2001. "The Chicago School of Ethnography." In *Handbook of Ethnography*, edited by Paul Atkinson, Amanda Coffey, Sara Delamont, John Lofland and Lyn Lofland, 11–25. Los Angeles, CA: Sage Publications.
Delacoste, Frédérique , and Priscilla Alexander, eds. 1998. *Sex Work: Writings by Women in the Sex Industry*. 2 ed. San Francisco: Cleis Press.
DeMichele, Matthew T. and Richard Tewksbury. 2004. "Sociological Explorations of Site-specific Social Control: The Role of the Strip Club Bouncer." *Deviant Behavior* 25:537–559.
Demovic, Angela. 1993. "Strippers in New Orleans: A Preliminary Study." *Human Mosaic* 27 (1–2):19–28.
DeNavas-Walt, Carmen, Bernadette D. Proctor, and and Cheryl Hill Lee. 2005. "Income, Poverty, and Health Insurance Coverage in the United States: 2004." *U.S. Census Bureau Current Population Reports*, 60-229. Washington, DC: U.S. Government Printing Office
Desmon, Jane. 1999. "Let's Lu'au." In *Staging Tourism: Bodies on Display from Waikiki to Seaworld*. University of Chicago Press.
Dewey, Susan. 2011. *Neon Wasteland: On Love, Motherhood, and Sex Work in a Rust Belt Town*. Berkeley: University of California Press.
Dewey, Susan, and Tonia St. Germain. 2014. "Sex workers/sex offenders: Exclusionary criminal justice practices in New Orleans." *Feminist Criminology* 10(3), 211–234.
Dietler, Michael and Brian Hayden, eds. 2001. *Feasts: Archaeological and Ethnographic Perspectives on Food, Politics, and Power*. Washington, DC: Smithsonian.
Di Giovine, Michael A. 2009. *The Heritage-scape: UNESCO, World Heritage, and Tourism*. Lanham, MD: Lexington Books.
Di Giovine, Michael A. 2014. "The Imaginaire Dialectic and the Refashioning of Pietrelcina." In *Tourism Imaginaries: Anthropological Approaches*, edited by Noel B. Salazar and Nelson H. H. Graburn, 147–171. New York: Berghahn.
Di Giovine, Michael A. and Ronda L. Brulotte. 2016. "Introduction: Food and Foodways as Cultural Heritage." In *Edible Identities: Food as Cultural Heritage*, edited by Ronda L. Brulotte and Michael A. Di Giovine, 1–28. New York: Routledge.
Dimanche, Frédéric. 2003. "The Louisiana Tax Free Shopping Program for International Visitors: A Case Study." *Journal of Travel Research*, 41:311–314.
Doezema, Jo. 1998. "Forced to Choose: beyond the voluntary v. forced prostitution dichotomy." In *Global Sex Workers: rights, resistance, and redefinition*, edited by Kamala Kempadoo and Jo Doezema, 34–50. New York: Routledge.
Donlon, J. G., and J. F. Agrusa. 2003. "Attraction of the Naughty-Gentleman's Clubs as a Tourism Resource: The French Quarter Example." In *Sex and Tourism: Journeys of Romance, Love, and Lust*, edited by T. G. Bauer and B. Mckercher, 119–136. New York: The Haworth Hospitality Press.
Douglas, Mary. 1987. *Constructive Drinking: Perspectives on Drink from Anthropology*. Mary Douglas Collected Works, Volume X. New York, NY: Routledge.
Dworkin, Andrea. 1981. *Pornography: Men Possessing Women*. United States: Putnam.
Egan, R. Danielle. 2004. "Eyeing the Scene: The Uses and (Re)uses of Surveillance Cameras in an Exotic Dance Club." *Critical Sociology* 30:299–319.

Egan, R. Danielle. 2006. *Dancing for Dollars and Paying for Love: The Relationships Between Exotic Dancers and Their Regulars*. New York: Palgrave Macmillan.
Enloe, C. 1990. *Bananas, beaches and bases: Making feminist sense of international politics*. Oakland: University of California Press.
Farmer, Paul E., Bruce Nizeye, Sara Stulac, and Salmaan Keshavjee. 2006. "Structural Violence and Clinical Medicineq." *PLoS Med* 3 (10).
Ferris, Jaqueline. 1997. "Courtship, drinking and control: A qualitative analysis of men and women's experiences." *Contemporary Drug Problems* 24 (4).
Foucault, Michael. 1975. *Discipline and Punish: The Birth of the Prison*. New York: Random House.
Fox, Robin. *Food and Eating: An Anthropological Perspective*. Social Issues Research Centre. http://www.sirc.org/publik/food_and_eating_0.html.
Frank, Katherine. 2002. *G-Strings and Sympathy: Strip Club Regulars and Male Desire*. Durham, NC: Duke University Press.
Frohlick, Susan. 2007. "Fluid Exchanges: The Negotiation of Intimacy between Tourist Women and Local Men in a Transnational Town in Caribbean Costa Rica." *City and Society*, 19(1) 139-168.
Frohlick, Susan. 2013. "Intimate Tourism Markets: Money, Gender, and the Complexity of Erotic Exchange in a Costa Rican Carribean Town." *Anthropological Quarterly* 86 (1):133–162.
Frow, John. 1991. "Tourism and the Semiotics of Nostalgia." *October by MIT Press* 57 (Summer):123–151.
Gilbert, Dennis. 2010. *The American Class Structure in an Age of Growing Inequality. 8th edition.* Thousand Oaks, CA: Pine Forge Press.
Goffman, Erving. 1959. *The Presentation of Self In Everyday Life*. Garden City, NY: Doubleday.
Goffman, Erving. 1963. *Stigma: Notes on the Management of a Spoiled Identity*. Englewood Cliffs, NJ: Prentice-Hall.
Goffman, Erving. 1974. *Frame Analysis: An Essay on the Organization of Experience*. Cambridge: Harvard University Press
Gotham, Kevin Fox. 2007. *Authentic New Orleans: Tourism, Culture, and Race in the Big Easy*. New York: New York University Press.
Graburn, Nelson. 1995. "Tourism, Modernity, and Nostalgia." In *The Future of Anthropology*, edited by Akbar Ahmed and Cris Shore. London: The Athone Press, 158–178.
Gravari-Barbas, Maria and Nelson Graburn. 2012. "Tourist Imaginaries." *Via@ Tourist imaginaries*, no. 1, 1–4. posted on March 16th, 2012, https://viatourismreview.com/wp-content/uploads/2015/06/Editorial1_EN-3.pdf.
Gursoy, Dogan, Yasin Boylu, and Umut Avci. 2011. "Identifying the Complex Relationships Among Emotional Labor and Its Correlatate." *International Journal of Hospitality Management*. 30 (4): 783–794. https://doi.org/10.1016/j.ijhm.2010.10.009
Gusfield, J. R. 1987. "Passage to Play: Rituals of Drinking Time in American Society." In *Constructive Drinking: Perspectives on Drink from Anthropology*, edited by Mary Douglas. Cambridge: Cambridge University Press.
Gusfield, Joseph R. 1996. *Contested Meanings: The Construction of Alcohol Problems*. Madison: University of Wisconsin Press.
Hacking, Ian. 1986. "Making Up People." In *Reconstructing Individualism: Autonomy, Individuality, and The Self in Western Thought*, edited by Thomas C. Heller, Sosna Morton and David E. Wellbery, 222-236. Stanford, CA: Stanford University Press.
Haddican, James. 1985. "Picayunes." *Times-Picayune*. July 6, 1985.
Hanna, Judith L. 1998. "Undressing the First Amendment and Corsetting the Striptease Dancer." *The Drama Review* 42 (2):32.
Hanna, Judith L. 2008. "Right To Dance: Exotic Dancing in the US." In *Dance, Human Rights, and Social Justice: Dignity in Motion*, edited by Naomi Jackson and Toni Shapiro-Phim. Lanham, MD: Scarecrow Press.
Hanna, Judith L. 2012. *Naked Truth: Strip Clubs, Democracy, and a Christian Right*. Austin, TX: University of Texas Press.

Harrison, Julia D. 2003. *Being a Tourist: Finding Meaning in Pleasure Travel.* Vancouver, BC: UBC Press.

Heath, Dwight B. 1987. "A Decade of Development in the Anthropological Study of Alcohol Use, 1970–1980." In Mary Douglas Collected Workes, Volume X, *Contructive Drinking: Perspectives on Drink from Anthropology*. New York, NY: Routledge.

Hennig-Thurau, Thorsten, Markus Groth, Michael Paul, and Swayne D. Gremler. 2006. "Are All Smiles Created Equal? How Emotional Contagion and Emotional Labor Affect Service Relationships." *Journal of Marketing* 70(3), pp. 58–73.

Herbert, F. Edward. 1974. "*Charlie the Mole and Other Droll Souls*. [Review of the book by the same title, by H. Jacobs]." *Louisiana History: The Journal of Louisiana Historical Association* 15 (1):89–90.

Holmila, Marja and Kirsimarja Raitasalo. 2005. "Gender Differences in Drinking: Why Do They Still Exist?" *Addiction* 100 (12):1763–1769.

Hopper, Hedda. 1948. "Looking at Hollywood: Lansbury May Be Lead." *Times-Picayune*, March 26, 1948, 24.

Hoschild, Arlie R. 1983. *The Managed Heart: The Commercialization of Human Feeling.* Berkeley: University of California Press.

Interview with the Vampire: The Vampire Chronicles. 1994. Neil Jordan, director. Geffen Pictures.

Jacobs, H. 1948. "Remoulade: Meets a Cop." *Times-Picayune*, December 26, 1948, 4, 2.

Jacobs, Howard. 1955. "B-Drink Sorority Imports Gendarme-Baffling Device." *Time-Picayune, Remoulade*. November 3, 1955.

Jacobs, H. 1973. *Charlie the Mole and Other Droll Souls*. Gretna: Pelican Publishing Company Inc.

Jacobs, H. 1975. "Berserk Reveler Is Thought Floor Show." *Times-Picayune*, September 15, 1975, 13, 1.

Janet. *Rocket Queen.* Independently published "zine," catalogued at the Newcomb Archives for Research on Women, Tulane University, New Orleans, LA.

Jex, Steve M. and Thomas W. Britt. 2014. *Organizational Psychology: A Scientist-Practitioner Approach*. Hoboken, NJ: John Wiley & Sons, Inc.

Joy, Charlotte. 2016. *The Politics of Heritage Management in Mali: From UNESCO to Djenné*. New York, NY: Routledge.

Katz, Jack. 1997. "Ethnography's Warrants." *Sociological Methods and Research* 25 (4):391–423.

Kempadoo, Kamala. 2004. *Sexing the Carribean: Gender, race, and sexual labor*. New York: Routledge.

Kempadoo, Kamala, and Jo Doezema. 1998. *Global Sex Workers: Rights, Resistance, and Redefinition*. New York, NY: Routledge.

Kempadoo, Kamala. 1998. "The Exotic Dancers Alliance: An Interview with Dawn Passar and Johanna Breyer." In *Global Sex Workers: Rights, Resistance, and Redefinition*. New York, NY: Routledge.

Kirshenblatt-Gimblett, Barbara. 1998. *Destination Culture: Tourism, Museums, and Heritage*. Berkeley, CA: University of California Press.

Klein, Marty 2012. *America's War on Sex: The Continuning Attack on Law, Lust and Liberty*. 2 ed. Santa Barbara CA: Praeger.

Ku, Robert Ji-Song 2013. *Dubious Gastronomy: The Cultural Politics of Eating Asian in the USA*. Honolulu: University of Hawaii Press.

Kulick, Don. 1998. *Travesti: Sex, Gender and Culture among Brazilian Transgendered Prostitutes*. Chicago: University of Chicago Press.

Lavin, Melissa F. 2014. "If You Want It, You Can Get It Right Here." *Humanity & Society* 38(2): 132–157.

Lee, V. 1970. "B-Drinking It's Difficult to Pin Down Reason Activity Continues: Officials Don't See Eye-to-Eye on Issue." *Times-Picayune*, October 19, 1970, 6, 2.

Leigh, Carol. 1997. "Inventing Sex Work." In *Whores and Other Feminists*. Jill Nagel, ed. New York: Routledge.

Levi-Strauss, Claude. 1969. *The Elementary Structures of Kinship*. Translated by James Harle Bell, John Richard von Sturmer, and Rodney Needham. Boston: Beacon Press.

Lewis, Peirce F. 1976. *New Orleans—The Making of an Urban Landscape*. Cambridge, MA: Ballinger Publishing Company.

Leite, Naomi. 2016. "Notes on Contributors." In *Tourism Imaginaries: Anthropological Approaches*. Noel B. Salazar and Nelson H. H. Graburn, eds. New York, NY: Berghahn.

Linz, Daniel; Kenneth C. Land; J. R. Williams; Bryant Paul; Michael E. Ezell. 2004. "An Examination of the Assumption That Adult Businesses are Associated with Crime in Surrounding Areas: A Secondary Effects Study in Charlotte, North Carolina." *Law and Society Review* 38 (1):69–104.

Lippard, Lucy. 1998. *The Lure of the Local*. New York: The New Press.

Little, Walter. 2003. "Performing Tourism: Maya Women's Strategies." *Signs: Journal of Women in Culture and Society* 29(2) 527–533.

Logan, Joshua, dir. 1956. *Bus Stop*. United States: Twentieth Century Fox.

Louisiana Department of Culture, Recreation & Tourism. *2006-7* Marketing Plan. Developed with New Orleans Marketing Corporation and the New Orleans Metropolitan Convention & Visitors Bureau.

Lowrance, Dee. 1941. "Cinema Cinderellas." *Times-Picayune*, November 16, 1941.

Long, Alecia P. 2005. *The Great Southern Babylon: Sex, Race, and Respectability in New Orleans, 1895–1920*. Baton Rouge: Louisiana State University Press.

MacCannell, Dean. 1976. *The Tourist, A New Theory of the Leasure Class*. New York: Schocken Books.

Mackinnon, Catharine. 1989. *Toward a Feminist Theory of the State*. Cambridge, MA: Harvard University Press.

Mackinnon, Catharine. 1993. *Only Words*. Cambridge, Massachusetts: Harvard University Press.

Mahane, T. P. 2012a. "Crime-Fighting Reporter Clifton Monroe Returns for Another Screwball Murder Mystery." *Times-Picayune*, Oct. 9. Http://blog.nola.com/entertainment_impact_arts/print.html?entry=/2012/10/crime-fighting_reporter_clifto.html.

Mahane, T. P. 2012b. "Hard-Boiled, Over Easy—Second Installment of Radio Drama Blends Comedy and Suspense." *Times-Picayune*, October 12, 2012, A9.

Malle, Louis, dir. 1978. *Pretty Baby*. Los Angeles, CA: Paramount Pictures.

The Man with the Golden Arm. 1955. Otto Preminger, director.

Manderson, L., and M. Jolly, eds. 1997. *Sites of Desire/ Economies of Pleasure*. Chicago: University of Chicago Press.

McKercher, Bob and Thomas G. Bauer. 2003. "Conceptual Framework of the Nexus Between Tourism, Romance, and Sex." In *Sex and Tourism: Journeys of Romance, Love, and Lust*, edited by Thomas G. Bauer and Bob McKercher, 3-18. Binghamton, New York: The Haworth Hospitality Press.

Melancon, Alex. 1941. "Juke Boxes, Dance Girls Invade Bayou County, Bee-Drinker Relates Tale of Her Hitch-Hiking Odyssey." *Times-Picayune*, February 2, 1941, 4, 2.

Metro-Roland, Michelle M. 2011. "Landscape Tourism." In *Tourists, signs and the city: the semiotics of culture in an urban landscape*. Farnham, UK: Ashgate Publishing.

Milner, Lauren E. 2012. *Respectably Dull: Striptease, Tourism and Reform in Postwar New Orleans*. M.A., History, University of New Orleans.

Minor, W. F. 1968. "Gov. Williams Blasts Guice." *Times-Picayune*, October 10, 1968, 14, 1.

Mintz, Sidney W. and Christine M. Du Bois. 2002. The Anthropology of Food and Eating. *Annual Review of Anthropology* 31: 99–119 https://doi.org/10.1146/annurev.anthro.32.032702.131011.

Mohr, Hubert. 2011. "Reflections on 'Museality' as a Critical Term in the Aesthetics of Religion." *Journal of Religion in Europe*, 4: 14–39.

National Center for Transgender Equality. 2014. *Transgender Terminology*. http://www.transequality.org/issues/resources/transgender-terminology.

National Restaurant Association. "Employee turnover rate tops 70% in 2015." *National Restaurant Association News & Research*, March 22,2016. http://www.restaurant.org/News-Research/News/Employee-turnover-rate-tops-70-in-2015.

New Orleans and Company Michie City Publications. 1956. *The Code of the city of New Orleans, Louisiana, 1956; General ordinances of the city. Ordinance No. 828 M. C. S.; adopted October 18, 1956, effective December 31, 1956. Published by order of the City Council.* New Orleans: Hauser Print Co.

New Orleans Paddlewheels, Inc. "The Paddlewheeler Creole Queen." http://www.creolequeen.com/creole-queen.html. Accessed 10/25/2017.

Olds, Kirsten. 2013. "Gay Life Artists: Les Petites Bonbons and Camp Performativity in the 1970s." *ArtJournal* 72(2) 16–33.

Paul, Bryant, Daniel Linz, and Bradley J. Shafer 2001 "Government Regulations of Adult Businesses Through Zoning and Anti-Nudity Ordinances: Debinking the Legal Myth of Negative Secondary Effects " *Communication Law and Policy* 6 (2):355–391.

Perales, Nan. 1983. "Ban on 'B-drinking' Struck Down." *Times-Picayune*. June 3, 1983.

Player, Eliza. 2010. *Heroin, Hurricane Katrina and the Howling Within*. Copyright Eliza Player, Kindle Books.

Price, Joshua M. 2012. *Structural Violence: Hidden Brutality in the Lives of Women*. Albany, New York: State University of New York Press.

Query, Julia, John Montoya, and Vicky Funari. 2001. *Live Nude Girls, UNITE!* Brooklyn, New York: First Run/ Icarus Films.

Regular Session of the Legislature of Louisiana, Act 360.

Renton vs. Playtime Theatres, Inc. 1986. 475 U.S. 41. No. 84–1360. Supreme Court of the United States.

Reinarman, Craig. 1988. "Prolegomenon to a Qualitative, Exploratory study of alcohol and sexuality in the context of AIDS (Working Paper)." *Alcohol Research Group Berkeley California* Working Paper.

Rice, Anne. 1976. *Interview with the Vampire*. New York: Alfred A. Knopf.

Roesler, B. 1985. "Rozelle Devotee of Horse Racing." *Times-Picayune*, April 20, 1985, C1.

Rood, Justin. 2008. "Sex Trade Spikes During Conventions." *ABCNews*, Aug. 22, 2008. http://abcnews.go.com/Blotter/story?id=5629167&page=1.

Rose, A. 1974. *Storyville, New Orleans: Being an Authentic, Illustrated Account of the Notorious Red-Light District*. Tuscaloosa: University of Alabama Press.

Sakakeeny, Matt. 2013. *Roll With It: Brass Bands in the Streets of New Orleans*. Durham, NC: Duke University Press.

Salazar, Noel. 2009. "Imaged or Imagined? Cultural Representations and the "Tourismification" of Peoples and Places." *Cahiers d'Etudes Africaines*, n193–194, p 49–72.

Salazar, Noel B. and Nelson H. H. Graburn, eds. 2014. *Tourism Imaginaries: Anthropological Approaches*. New York: Berghahn.

Sawyer, Charles and Capt, J. C. 1948. *Current Population Reports Consumer Income*. Edited by Department of Commerce. Washington, DC.

Schlosser, Eric. 2003. *Reefer Madness: Sex, Drugs, and Cheap Labor in the American Black Market*. New York: Houghton Mifflin Company.

Sebastian, Rachael. 2015. *If You Like Our Sound, Stick Around: Buskers of the French Quarter*. MA thesis, Wichita State University.

Shelton, R., dir. 1989. *Blaze*. United States: Buena Vista Pictures.

Shteir, Rachel. 2004. *Striptease: The Untold History of the Girlie Show*. New York: Oxford University Press.

Simoni, Valerio. 2015. "Breadwinners, Sex Machines and Romantic Lovers: Entangling Masculinities, Moralities, and Pragmatic Concerns in Touristic Cuba." *Etnográfica* 19 (2), 389–411.

Smith, Valene. 1989. *Hosts and Guests*. Philadelphia: University of Pennsylvania Press.

Social Issues Research Centre. 1998. *Social and Cultural Aspects of Drinking: A Report to the European Commission*. Oxford: The Social Issues Research Centre.

Sontag, Susan. 1966. "Notes on Camp." *Against Interpretation: And Other Essays*, 275–92. New York: Farrar, Straus & Giroux.

Starr, B. , and H. Perry. 1974. *Blaze Starr: My Life as Told to Huey Perry*. New York: Praeger.

Stewart, Susan. 1994. *Crimes of Writing: Problems in the Containment of Representation*. Durham, NC: Duke University Press.

Sundahl, Debi. 1998. "Stripper." In *Sex Work: Writings by Women in the Sex Industry*, 2nd edition. Edited by Frédérique Delacoste and Priscilla Alexander. San Fransisco: Cleis Press, 175–180.
Toole, John Kennedy. 1980. *A Confederacy of Dunces*: Baton Rouge: Louisiana State University Press.
Turner Classic Movies, Inc. 2016. *Among the Living*. http://www.tcm.com/tcmdb/title/67278/Among-the-Living/.
Turner, Victor. 1969. *The Ritual Process: Structure and Anti-Structure*. Ithaca, New York: Cornell University Press.
Turner, Victor. 1974. *Dramas, Fields, and Metaphors: Symbolic Action in Human Society*. Cornell University Press.
Urry, John. 1990. *The Tourist Gaze: Leisure and Travel in Contemporary Societies*. London: Sage.
Urry, John and Jonas Larsen. 2011. *The Tourist Gaze 3.0*. Sage. Los Angeles, California: Sage.
Van Maanen, J. and Schein, E. H. 1979 "Toward a theory of organizational socialization." *Research in Organizational Behavior*, 1, 209–264.
Vesey, C. and Dimanche, F. 2003. "From Storyville to Bourbon Street: Vice, Nostalgia and Tourism." *The Journal of Tourism and Culture Chance* 1 (1):54–70.
Waller, Richard. 1993. "Conclusions." In *Being Maasai: Ethnicity and Identity in East Africa*, edited by Thomas Spear and Richard Waller, 290-302. London: James Currey.
Wardlow, Holly. 2006. *Wayward Women: Sexuality and Agency in a New Guinea Society*. Berkeley, CA: University of California Press.
Waterton, Emma and Steve Watson. 2014. *The Semiotics of Heritage Tourism*. Channel View Publications.
Weber, Max. 1958. *The Protestant Ethic and the Spirit of Capitalism*. New York: Scribner.
White, Louise. 1990. *The Comforts of Home; Prostitution in Colonial Nairobi*. Chicago: University of Chicago Press.
Whyte, Martin K. 1990. *Dating, Mating, and Marriage*. New York: Aldine de Gruyter.
Williams, T. Harry. 1969. *Huey Long*. New York: Alfred A Knopf, Inc.
Williams, Erica Lorraine. 2013. *Sex Tourism in Bahia*. Urbana, Chicago, and Springfield: University of Illinois.
Williams, John, Maria Ortiz, Marius Mihai, and Karen B. Brakel. 2013. *University of New Orleans Division of Business and Economics Research Metropolitan Report: Economic Indicators for the 4th Quarter 2013 through the 3rd Quarter 2015*. New Orleans: University of New Orleans.
Zemeckis, Leslie. 2013. *Behind the Burly Q: The Story of Burlesque in America*. New York: Skyhorse Publishing, Inc.

Index

About the Author

Angela R. Demovic is a sociocultural anthropologist who received her PhD from Tulane University. She has taught anthropology and women's studies at Tulane University, Louisiana State University, and Wichita State University. Her research focuses on gender, political economy, and tourism.

www.ingramcontent.com/pod-product-compliance
Ingram Content Group UK Ltd.
Pitfield, Milton Keynes, MK11 3LW, UK
UKHW020210290926
15470UKWH00001B/100

9 781498 531320